CHEMISTRY OF
MARINE NATURAL PRODUCTS

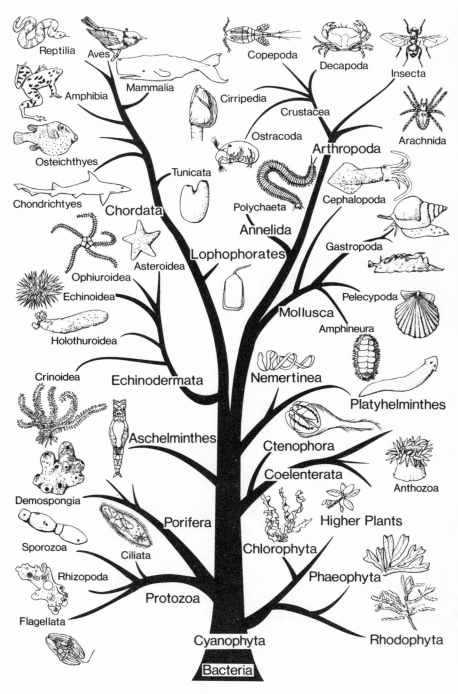

The phylogenetic tree.

Chemistry of Marine Natural Products

PAUL J. SCHEUER

Department of Chemistry
University of Hawaii
Honolulu, Hawaii

ACADEMIC PRESS   1973   *New York and London*

ACADEMIC PRESS, INC.
111 Fifth Avenue, New York, New York 10003

United Kingdom Edition published by
ACADEMIC PRESS, INC. (LONDON) LTD.
24/28 Oval Road, London NW1

LIBRARY OF CONGRESS CATALOG CARD NUMBER: 72-88326

PRINTED IN THE UNITED STATES OF AMERICA

To Alice

CONTENTS

PREFACE

Recent publication of Halstead's three-volume treatise, "Poisonous and Venomous Marine Animals," of Baslow's, "Marine Pharmacology," and of Martin's two-volume "Marine Chemistry" attests to a heightened interest in the marine environment by the scientific community. Many factors have contributed to this surge of activity, but two are perhaps of critical importance to the subject matter of this book. One of the factors lies in the vastly improved accessibility of many ocean areas of the world. Second, our research tools of the past ten years or so have enabled us to carry out separations of complex mixtures, to perform structural determinations on tiny amounts of material, and to do so much more rapidly than was the case even fifteen years ago. Interest in studying and in preserving the marine environment is particularly great among today's high school and college-age population. If this interest is to be translated into a lasting impact by kindling new scholarship, today's teachers need the necessary resources to sustain their students' interest. In the area of structural organic chemistry pertaining to the marine environment, no monograph currently in print gathers and evaluates the state of our knowledge in this field. There are a few highly specialized short review articles, but none of a comprehensive nature. As a modest contributor over the years to the review literature of limited scope, I have come to the conclusion that a genuine gap exists in the chemical literature dealing with marine natural products which I hoped to bridge with this book.

Chemists with an interest in the sea comprise the primary group of people for whom this book is intended. They can be graduate students, upper division undergraduates, or teachers. This work should also appeal to marine biologists, oceanographers, pharmacologists, and their students. It can be the basis of a special topics course or can be used for supplementary reading in a course dealing with structural organic chemistry. Hopefully, there will be some people who will read it mostly for pleasure.

Historically, all organic chemistry was the chemistry of naturally occurring carbon compounds. Over the years, natural product chemists devoted themselves to the study of what we now often refer to as secondary metabolites. Some natural products, among them amino acids, carbohydrates, and their polymers, were recognized as the universal constituents of living matter and were considered to be primary metabolites. They became the domain of biochemistry.

Quantitatively, the number of known naturally occurring organic compounds is vastly outdistanced by the number of known synthetic organic compounds. Economically, organic natural products of commerce (sucrose, quinine, antibiotics notwithstanding) are no match in value for the tremendous empire that has resulted from organic synthesis (solvents, dyes, monomers for fibers, rubbers, plastics, drugs). Yet natural products continue to hold a unique fascination not only because of their relationships to the organisms from which they are derived, not only because plants that can photosynthesize possess synthetic power that far surpasses that of man, not only because of their potential usefulness to man in their natural form or as templates for synthetic analogs, but chiefly because each new structural type reveals something of nature's molecular architecture and poses new questions of how and why these compounds are being produced. Since few natural products have been isolated so far from marine organisms, it is not yet possible—except in a few isolated areas that have been well studied—to make intelligent predictions about the likely outcome of an investigation, a feature that makes this area of natural products research particularly attractive.

The time has long passed since a book on chemistry—no matter how narrow its scope—could hope to offer an exhaustive treatment of the subject. This book makes no such attempt. My major emphasis has been structural organic chemistry. In this context I have tried to exclude active principles or factors and functional types whose structures are incompletely known. I have attempted to include structural formulas of even relatively simple organic compounds because it has been my experience in recent years that nomenclature of organic compounds, particularly that of many time-honored trivial names, is no longer emphasized in undergraduate curricula. As a result, many potential readers would have to consult an organic chemical dictionary with considerable frequency if fewer structural formulas were presented. Furthermore, I have attempted to minimize the use of a single formula for an entire series of compounds because of my own dislike of this practice.

From the time when I began to think about the organization of this book I was convinced that it should be written along chemical lines. The time has come to emphasize structural relationships, and it may be advantageous to view marine taxa from the up-to-now unconventional angle of structural chemistry. This approach should help to focus attention on the peaks and troughs that exist in our knowledge of marine natural products.

The only meaningful approach for a book dealing with marine natural products is one that follows broad biogenetic concepts rather than functional group concepts. This is true even though practically none of the biosynthetic relationships of marine-derived organic compounds have been experimentally verified. Our considerable body of knowledge of biogenesis of secondary metabolites in terrestrial organisms must therefore serve as a guideline unless shown to be incorrect by future experimentation. Although it is reasonable to assume that few new and unique fundamental biogenetic pathways are apt to emerge from future marine research, even though the metabolic end products may have new and unique structures, questions such as the halogenation mechanism (why so few chlorine and so many bromine compounds?), the synthesis of the sterol side chain of gorgosterol and its congeners, or the synthesis of some of the complex marine toxins are fascinating topics for future research.

I have frequently attempted to point out where a particular plant or animal "belongs" in its genealogy. In addition, a phylogenetic tree is included which highlights the biological relationships and which hopefully serves as a ready reference for the reader.

Although there are a few cross references in the book, each chapter stands largely on its own and can be read independently of the others. Tables at the end of each chapter or major section list the compounds and provide key references.

It is a pleasure to acknowledge the help of my colleagues and associates who have contributed greatly to the completion of this work. C. W. J. Chang, J. S. Grossert, D. R. Idler, R. E. Moore, and A. W. Sangster read one or more of the chapters and made valuable comments and suggestions. My long-term collaborator A. H. Banner provided the sketch of the phylogenetic tree on which Susan G. Monden based the design and illustration. Mavis Kadooka typed the entire manuscript and drew many of the formulas.

Finally, I should like to express my gratitude to the members of my family, my students, my colleagues, and all others who patiently suffered my neglect while I was preparing the manuscript.

PAUL J. SCHEUER

1

ISOPRENOIDS

Isoprenoids, perhaps more than any other class of organic compounds, are associated in the minds of many biological and physical scientists with the chemistry of natural products. These compounds therefore form a suitable beginning for our discussion.

Wallach (1887) first recorded the remarkable observation that a distinct group of naturally occurring substances may be formally derived from branched five-carbon units. The ensuing theory that these compounds, the terpenes or perhaps more accurately the terpenoids, are biosynthesized from branched five-carbon modules was first stated by Ruzicka and Stoll (1922). Ruzicka's theory has been one of the most fruitful in organic chemistry. It has inspired and guided an incredible amount of imaginative research in structural determination and in organic synthesis—*in vitro* and *in vivo*. Beyond that, much research in physical-organic chemistry, in the area of molecular rearrangements in particular, has been sparked and sustained by the monoterpenes of the camphane series. In their own right, the terpenoids have long mirrored the development of organic chemistry. At the outset, as extractives of terrestrial plants, the terpenoids were prized for their pleasant aroma. For many years they were an interesting subject for academic research. More recently, with the advent of sophisticated methods of isolation, separation, and structural elucidation, terpene research has expanded greatly and has established these compounds as one of the most diverse and intriguing groups of natural products. Little wonder then that isoprenoid substances have been isolated not only from their traditional substrates, essential oils of vegetable

1

origin, but also from such varied natural sources as microorganisms, insects, and marine plants and animals.

The standard division of the terpenes according to carbon content into mono-(C_{10}), sesqui-(C_{15}), di-(C_{20}), etc., terpenes is logical and convenient and will be followed in this chapter. Because of our fragmentary exploration of the marine environment to date, it is impossible to tell whether the distribution of various isoprenoids in marine organisms parallels that found on land. A few ubiquitous monoterpenes, e.g., geraniol, α-pinene, have been reported from algal sources. The bulk of that work was done in the 1950's by Katayama, who later (1962) summarized it. The largest, yet still very modest, number of terpenoids from marine sources have been sesquiterpenoids; there have been a few diterpenoids, including some C_{19} and C_{21} offshoots; and a number of rather closely related triterpenoids of unique structure have been reported in the literature. Surprisingly, only two widely distributed triterpenoids, friedelin and taraxerol, have so far been isolated from marine sources (Tsuda and Sakai, 1960; Santos and Doty, 1971). And last, we have some unique representatives of the tetraterpenoid carotenoid pigments. Once the extent of our knowledge of marine terpenoid compounds approaches that of terrestrially derived substances it will be fruitful to compare distribution and oxidation patterns, and to relate these findings to other aspects of evolutionary theory. In this connection, the fascinating observation (*vide infra*) may be mentioned that a number of marine sesquiterpenes are optical antipodes of the corresponding terrestrial compounds.

A. Sesquiterpenoids

In a recent comprehensive review of sesquiterpenoid chemistry by Bryant (1969) the author states that through 1964 more than 200 sesquiterpenoids, divisible into 40 types, had been characterized. These figures—an average of five compounds per structural type—are a good indicator of the great versatility of these compounds, which are made up of only fifteen carbon atoms. Since the number of marine sesquiterpenoids that have been characterized through 1971 totals fewer than 40 compounds, a division of this section by individual structural type would be an exercise in overorganization.

Instead the presentation will be a pragmatic one. A brief discussion of previously known compounds will be followed by an account of structures which so far appear to be and perhaps are characteristic of the marine environment. So far, sesquiterpenoids have been isolated from only four marine phyla, the coelenterates (Cnidaria), the mollusks (Mollusca), the brown algae (Phaeophyta), and the red algae (Rhodophyta). Clearly, therefore, an attempt to assess phylogenetic significance is premature.

1. PREVIOUSLY KNOWN COMPOUNDS

One of the earliest reports of marine-derived sesquiterpenoids was published by Takaoka and Ando (1951). In this rarely quoted study the Japanese authors steam-distilled the brown alga *Dictyopteris divaricata* and obtained an oil with a "beach odor." The oil was said to consist of sesquiterpene hydrocarbons and alcohols as evidenced by the azulene which was obtained on dehydrogenation. One of the alcohols was designated as "1-cadinol," mp 137.5°–138.5°. This work was performed before sesquiterpene research had become attractive through efficient separation methods and non-destructive tools for structural determination. Irie and co-workers (1964) reexamined this research and confirmed, *inter alia*, that the sesquiterpene alcohol of Takaoka and Ando was (−)-δ-cadinol (1), known previously from the terrestrial genus *Pinus*.

By far the bulk of the sesquiterpenoids up to now have been isolated from the phylum Coelenterata, first reported on by Ciereszko *et al.* (1960). The phylum Coelenterata (Cnidaria) comprises some 9000 species in three classes: the Hydrozoa (e.g., the Portuguese man-of-war), the Scyphozoa (the jellyfish), and the Anthozoa (the sea anemones; soft and stony corals). The Anthozoa constitute the largest class (some 6000 species) and are divided into two subclasses: the Zoantharia (sea anemones and corals) and the Alcyonaria (sea fans, sea pansies, etc.). One of four orders of the alcyonarians are the Gorgonacea or gorgonians. All sesquiterpenes that have so far been reported from the marine coelenterates have been isolated from gorgonians, and all of them by one research group at the University of Oklahoma. Collection of these animals is an esthetic experience since gorgonians contribute some spectacularly beautiful species to tropical coral reefs.

In their first publication on terpenoid constituents of the gorgonians Ciereszko *et al.* (1960) reported the isolation of several partially characterized compounds in addition to a dextrorotatory cadinene from *Plexaura crassa*, which had been collected at Bermuda. The identity of this widely occurring sesquiterpene diene was secured by its conversion to a (+)-cadinene dihydrochloride and by selenium dehydrogenation to cadalene (1,6-dimethyl-4-isopropylnaphthalene). The infrared spectrum of the dihydrochloride was identical with that of a levorotatory sample prepared from oil of cade. The authors did not state which of the cadinene isomers was originally isolated. As Gallagher and co-workers (1964) pointed out, that might have been a difficult task without rigorous proof of homogeneity. In fact, the Australian workers further caution that cadinene dihydrohalides can arise not only from the nine cadinene isomers, but from other structures, e.g., copaene (*vide infra*, 14) as well.

Regardless of the correct structure of the sesquiterpene, its isolation constituted an authentic occurrence of a typical plant constituent in a marine animal, a carnivore at that! Being aware of the unusual nature of this finding Ciereszko *et al.* (1960) suggested that the terpenoids might be synthesized not by the gorgonians but by the unicellular algae, the zooxanthellae, which are symbiotically associated with the animals. Subsequently Ciereszko (1962) examined the zooxanthellae associated with the gorgonian *Pseudoplexaura crassa** and found that the algae alone contained at least 3% cadinene (based on dry weight) and 8% of a substance that was designated crassin acetate. Crassin acetate, a diterpene lactone, had also been isolated and partially characterized during the earlier survey (Ciereszko *et al.*, 1960).

A well-defined $(-)$-γ_1-cadinene (2) was reported by Irie *et al.* (1964) in their study of the brown alga *Dictyopteris divaricata*, along with $(-)$-δ-cadinol (1) and cadalene (1,6-dimethyl-4-isopropylnaphthalene), the latter presumably an artifact generated during isolation.

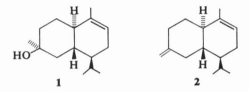

Closely related to the cadinene structure are the sesquiterpenes β-bisabolene (3), calamenene (4), and α-muurolene (5). All three compounds were isolated by the Oklahoma group (Weinheimer *et al.*, 1968) from gorgonians. All three compounds are dextrorotatory.

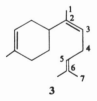

3

Bisabolene is widely distributed in terrestrial plants. Again, a number of double-bond isomers have been reported. The β-isomer designation, however, appears to be preferred for the 1,5-heptadiene structure (Manjarrez and Guzmán, 1966; Stevens, 1969) rather than for the 2,5-heptadiene as indicated in structure 3 according to Weinheimer *et al.* (1968b).

Calamenene (4) is relatively rare. A levorotatory, $[\alpha]_D^{26°} -68°$, isomer has

---

* Presumably *Pseudoplexaura crassa* is the same animal as the *Plexaura crassa* (*vide supra*) of Ciereszko's earlier publication (1960).

recently been isolated by de Mayo *et al.* (1965) from cedarwood of *Cedrela toona*. A semisynthetic sample, which was obtained by Joshi and co-workers (1964) as a result of aromatization of khushinol with *N*-lithioethylene-diamine, had a small positive rotation, $[\alpha]_D$ +0.88°. Apparently this sample represents a largely racemized substance. The small positive rotation may well be zero within the experimental error since the Indian workers do not specify whether rotations were determined visually or photoelectrically. Another semisynthetic sample which was prepared by Trivedi *et al.* (1966) from khusinol with boron trifluoride etherate, had a rotation of $[\alpha]_D$ +82.58°. The calamenene from the gorgonian had $[\alpha]_D^{25°}$ +55.4°.

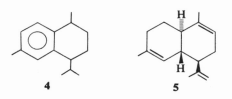

The third hydrocarbon in this group isolated from *Eunicea mammosa* is (+)-α-muurolene (**5**), $[\alpha]_D^{25°}$ +67.1° (Weinheimer *et al.*, 1968b). It is, except for rotation—and therefore configuration—identical with the terrestrially derived (−)-α-muurolene, $[\alpha]_D^{22°}$ −52.5° of Zabża and co-workers (1966).

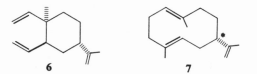

From the gorgonian *Eunicea mammosa* Weinheimer *et al.* (1970) isolated the related three compounds (+)-β-elemene (**6**), (−)-germacrene-A (**7**), and (−)-β-selinene (**8**). (+)-β-Elemene (**6**) had been obtained earlier from the same animal species which led to the authors' suspicion (Weinheimer *et al.*, 1968) that they might be dealing with an artifact and that the natural ses-quiterpene was of the germacrene type that underwent a Cope rearrangement during work-up. Using gentle isolation procedures, Weinheimer *et al.* (1970) isolated germacrene-A (**7**) and confirmed its ready isomerization to β-elemene (**6**) at elevated temperature or on silicic acid adsorbents. The degree of stereospecificity of this rearrangement poses an intriguing question that cannot be answered because of a lack of data. The gorgonian-derived β-ele-mene is dextrorotatory, $[\alpha]_D^{25°}$ +15.1°, as is the β-elemene isolated by Irie *et al.* (1964) from the brown alga *Dictyopteris divaricata* ($[\alpha]_D^{23°}$ +12.8°). Another dextrorotatory β-elemene, $[\alpha]_D^{20°}$ +14.2°, has been reported from a terrestrial source by Pigulovskii and Borokov (1962). However, material that constituted two original isolations from sweet flag (Šorm *et al.*, 1953) and juniper oils (Sykora *et al.*, 1956) by the Czech terpene group was levorotatory, $[\alpha]_D^{20°}$ −20° and $[\alpha]_D$ −16.9°, respectively. Unfortunately, no rotation is reported by Patil and Rao (1967) for a sample of β-elemene synthesized from an optically active bicyclic precursor. Since the natural precursor of β-elemene, germacrene-A (7), has one center of chirality (starred carbon atom), which is unaffected by the rearrangement, much of the observed optical activity of the various elemenes may be due to that carbon. Germacrene-A, $[\alpha]_D^{25°}$ −3°, has not been isolated from another source so far and assessment of its optical purity is not possible.

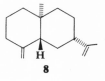

8

The third component in this group, β-selinene (8) would also appear to be an artifact. Weinheimer *et al.* (1970) designate it to be levorotatory, but report no specific rotation. The Russian workers (Pigulevskii and Borokov, 1962), who isolated β-selinene along with their dextrorotatory β-elemene, report $[\alpha]_D^{20°}$ −25.99°. Other reported rotations for β-selinene are $[\alpha]_D$ +32° for the isolated liquid and +63° for the sesquiterpene regenerated via the hydrochloride (Ruzicka *et al.*, 1931) and −46° from yet another source (Dixit *et al.*, 1967). Clearly, this group of sesquiterpenes bears reinvestigation. In particular, the germacrene–elemene rearrangement needs to be studied with reference to optically pure standards. Beyond that, the biosynthesis of antipodal terpenoids by different organisms is of considerable fundamental interest.

From another gorgonian, *Pseudopterogorgia americana*, Weinheimer and co-workers (1968c) isolated the related sesquiterpenes 9-aristolene (9),1 (10)-aristolene (10), and (+)-γ-maaliene (11). The two aristolenes, 9 and 10, have been well characterized from a variety of terrestrial sources, although not without the confusion that was associated with sesquiterpene chemistry until recently. The enantiomer of 9-aristolene depicted by 9 is known in the literature as α-ferulene, $[\alpha]_D$ +68°, isolated from the genus *Ferula* of the parsley family (Umbelliferae) by Carboni *et al.* (1965). The rotation of the compound isolated from the gorgonian (Weinheimer *et al.*, 1968) is $[\alpha]_D^{25°}$ +80.9°. A common synonym of the 1(10)-aristolene is β-gurjunene. The commonly occurring enantiomer is dextrorotatory (Streith *et al.*, 1963; Vrkoč *et al.*, 1964), while the marine-derived compound (**10**) has $[\alpha]_D^{25°} - 78.5°$. Compound **11**, γ-maaliene, $[\alpha]_D^{25°} + 10.9°$, had not been isolated from another natural source, but was known as a member of a mixture of olefinic isomers, intermediate in the synthesis of maaliol by Bates and co-workers (1960).

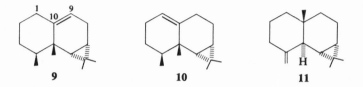

Three additional tricyclic sesquiterpenes, all previously known, have been reported from marine organisms, two from a gorgonian and one from a brown alga. The Oklahoma group (Weinheimer *et al.*, 1968) isolated from *Pseudoplexaura porosa* (+)-α-cubebene (**12**) and (+)-β-ylangene (**13**), while Irie *et al.* (1964) reported (−)-copaene (**14**) from *Dictyopteris divaricata*. Authentic α-cubebene, $[\alpha]_D^{30°} - 20°$, was first isolated only recently (Ohta *et al.*, 1966) from oil of cubeb and its structure was confirmed by synthesis in two laboratories (Tanaka *et al.*, 1969; Piers *et al.*, 1969). α-Cubebene, as isolated from the gorgonian (Weinheimer *et al.*, 1968b) is dextrorotatory, $[\alpha]_D^{25°} + 23.6°$.

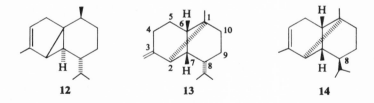

β-Ylangene (**13**), $[\alpha]_D^{25°} + 9.2°—10.6°$, is a rare sesquiterpene. Its only other isolation has been from the oil of the Valencia orange by Hunter and Brogden (1964) who reported no rotation data. It is readily isomerized to the much more common α-ylangene, which differs from the β-isomer by having a 3,4-double bond instead of the exocyclic one. It is possible if not likely that β-ylangene is much more common than is now assumed and that its ready isomerization accounts for its rarity. The stereochemistry of α-ylangene has been established by Ohta and Hirose (1969). It, in turn, is isomeric with α-copaene (**14**), from which it differs only with respect to configuration at C-8. In a recent isolation of α-copaene from oil of cubeb Ohta *et al.* (1968)

report a rotation of $[\alpha]_D$ −19°, close to the rotation of α-copaene, $[\alpha]_D^{23°}$ −26.1°, isolated from the brown alga (Irie *et al.*, 1964).

2. New Compounds

The sesquiterpenoids of novel structure fall into four well-defined categories. By far the most numerous group, based on a rearranged farnesyl skeleton, which so far encompasses about a dozen compounds, has been isolated by two Japanese and an American group from sea hares and from red algae of a single genus. Many of these compounds contain bromine, and two contain chlorine as well. Two other categories have only one representative each: one a benzofuran derivative and the other a rearranged maaliene (**11**), both isolated from gorgonians. The fourth group consists of two oxygenated derivatives of known skeletal type. These groups of compounds will be discussed in reverse order, beginning with those that are most closely related to known sesquiterpenoids.

Two functionalized derivatives, an alcohol and a ketone, of the well-known β-selinene (**8**) were isolated by Irie's group (Kurosawa *et al.*, 1966) from the essential oil of the brown alga *Dictyopteris divaricata*. The alcohol had been detected earlier (Irie *et al.*, 1964) and was shown to have the structure of selinen-1β-ol (**15**), an inseparable mixture of the two double-bond isomers α- (**15a**) and β- (**15b**) selinene, $[\alpha]_D$ −30.8°. It was named dictyopterol and its structure was secured by chemical and spectral means.

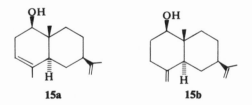

15a   **15b**

The second compound, β-selinen-1-one or dictyopterone, $[\alpha]_D$ −12.5°, was shown to have structure **16** by spectral data and chemical transformations.

The rearranged maaliene, β-gorgonene (**17**), was isolated by Weinheimer

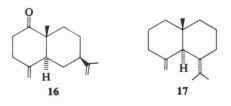

16   **17**

et al. (1968) from the gorgonian *Pseudopterogorgia americana*. It is the predominant sesquiterpene and occurs along with maaliene (**11**) and the aristolenes (**9, 10**). β-Gorgonene, $[\alpha]_D^{25°}$ +13.9°, was isolated as its crystalline silver nitrate, mp 132.5°–133.5°. The structure of this complex was determined by X-ray diffraction techniques (Houssain and Van der Helm, 1968). It has an interesting nonisoprenoid skeleton that may well have arisen from a common precursor with maaliene (**11**). As the authors (Weinheimer *et al.*, 1968c) point out, such a rearranged skeleton is analogous to the monoterpene sylvestrene (3-isopropenyl-1-methylcyclohexene), which is presumed to arise from carenes during isolation (Pollock and Stevens, 1965).

The benzofuran is furoventalene, a nonfarnesyl sesquiterpene of structure **18**. Its structure was determined by Weinheimer and Washecheck (1969)

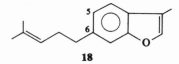

18

following its isolation from the sea fan *Gorgonia ventalina*. Analysis of spectral data failed to distinguish furoventalene (**18**) from its 5-alkylated isomer. Decision in favor of the 6-alkylated benzofuran was secured by synthesis of both compounds. The successful synthesis of furoventalene was accomplished by the following sequence. The aromatic starting moiety, 6-bromo-3-methyl-benzofuran (**20**), was the minor product of the cyclization of *m*-bromophenoxyacetone (**19**). The Grignard derivative of **20** was condensed with the ethylene ketal of levulinaldehyde (**21**). The resulting alcohol, when subjected to hydrogenolysis, yielded compound **22**, where hydrogenolysis of the benzyl alcohol, reduction of the furan ring, and ketal cleavage had taken place. Dehydrogenation regenerated the furan, which in turn was transformed to the corresponding tertiary alcohol **23** with methylmagnesium iodide. Dehydration of the alcohol led to the desired furoventalene **18** and other double-bond isomers, from which it was separated.

As the authors (Weinheimer and Washecheck, 1969) point out, it appears unlikely that this sesquiterpene is synthesized in nature from a farnesyl

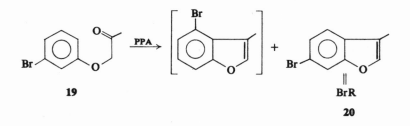

19   **BrR**

20

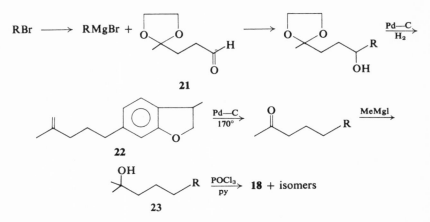

21

22

23

precursor. Instead it appears more likely that furoventalene is biosynthesized from a suitable monoterpenoid precursor, as e.g., piperitenone, **24**, by condensation with a five-carbon fragment.

24

The final group of sesquiterpenes to be considered are the interesting constituents of the gastropod mollusk *Aplysia kurodai* and of various species of red algae of the genus *Laurencia*. Undoubtedly all of these compounds are farnesyl-derived sesquiterpenes; however, only spirolaurenone (**45**) and the constituents of *L. pacifica* (**47** and **48**) represent an unrearranged farnesyl skeleton (**25b** and **25c**). In most of the compounds reported so far a C-11 methyl group has migrated to C-10 (**25e**) while some of the compounds are interesting intermediates (**25d**) in this rearrangement. No unrearranged compound with the cyclization pattern **25a** has so far been found in a marine source. Several of the six compounds possessing a **25d** skeleton are benzo-furans, but clearly unrelated to furoventalene (**18**). These were the first to be isolated from a species of sea hare, *Aplysia kurodai*, by Yamamura and Hirata (1963).

The sea hares belong to the large phylum Mollusca (mollusks). Over 80,000 living species have been described, which makes this phylum second in size to the arthropods. The phylum encompasses such familiar animals as clams, oysters, squids, and snails; as a consequence, it ranks in general popularity with birds and mammals. The largest class in the phylum are the gastropods (Gastropoda) comprising some 35,000 living and 15,000 fossil

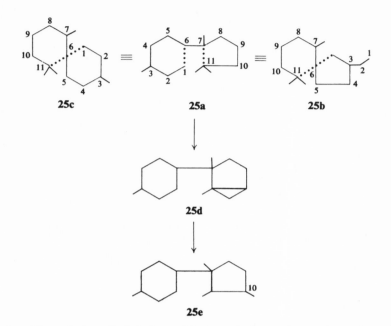

25c   25a   25b

25d

25e

species. The subclass Opisthobranchia is entirely marine and consists of the orders Tectibranchia (sea hares, bubbleshells), Pteropoda (sea butterflies), and Nudibranchia (sea slugs). The sea hares, which lack a shell, are unspectacular animals in color and shape and are easily missed among the rocks of a tidepool. Their trivial name stems from their appearance as they feed on a stand of algae, which is said to be reminiscent of terrestrial rabbits in a lettuce patch.

Yamamura and Hirata's (1963) work on the constituents of *Aplysia kurodai* followed an earlier inconclusive study by Tanaka and Toyama (1959). As part of an extensive investigation into the lipids of marine organisms these authors detected two crystalline bromine containing compounds in the unsaponifiable portion of *A. kurodai*. One of the compounds, mp 69°–71°, had the apparent composition $C_{16}H_{21}BrO$ and was characterized as a bromophenol; the second compound, mp 146.5°–147.5°, was insufficiently characterized.

Yamamura and Hirata (1963) followed up on the earlier work in an attempt to verify the occurrence of the bromo constituents of sea hares. They extracted whole dry animals with ether and saponified the extract with methanolic potassium hydroxide. The unsaponifiable matter was again extracted with ether, then hexane. The hexane-soluble residue was chromatographed first on silica gel, then on alumina. From the hexane fraction the authors isolated debromoaplysin (**26**) as an oil, which they converted to a mononitro derivative,

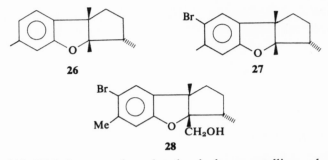

mp 107°–108°. With benzene they eluted colorless crystalline aplysin* (**27**), mp 85°–86°, $[\alpha]_D^{27°}$ −85.4°. The third constituent, aplysinol (**28**), emerged from the silica gel column by hexane–benzene elution as an oil. This oil when rechromatographed on alumina, was eluted with ether–methanol and yielded colorless crystals of aplysinol, mp 158°–160°, $[\alpha]_D^{19°}$ −55.6°.

Structures of the three compounds were determined by a combination of chemical degradation and spectral data. No stereochemistry was assigned, but on the basis of nmr data it was concluded that the hydroxymethylene and angular methyl groups of aplysinol (**28**) are cis to each other: The two angular methyl groups of aplysin (**27**) exhibit signals of δ1.25 and δ1.30, while in aplysinol (**28**) these two signals have been replaced by a two-proton singlet at δ3.78 (—CH_2OH) and a three-proton singlet at δ1.43, shifted downfield from the corresponding signals in aplysin (**27**).

Racemic aplysin (**27**) and debromoaplysin (**26**) were synthesized by Yamada *et al.* (1968), thereby confirming the structures of these two compounds. Aplysinol (**28**) had been related to the two aplysins earlier by Yamamura and Hirata (1963).

A closely related group of compounds as well as the three aplysins (**26, 27, 28**) have been isolated from red algae of the genus *Laurencia* (Rhodomelaceae) by Irie and co-workers and reported in a series of publications beginning in the mid-1960's. Actually, perhaps the first sesquiterpenoids from a *Laurencia* species were reported by Obata and Fukushi (1953). From an extract of the steam distillate of *L. glandulifera* these authors isolated fractions that they believed to be a sesquiterpene and a sesquiterpene alcohol. In fact Irie and co-workers (1965a) did not isolate a terpenoid compound at all in their first investigation into *Laurencia* constituents, but laurencin, which is an oxocin derivative of a nonisoprenoid precursor (see Chapter 5). However, in

---

* The name *aplysin*—no doubt Yamamura and Hirata (1963) were unaware of this— had already been used by Winkler *et al.* (1962) to designate the toxic extract of the digestive gland of *A. californica* and *A. vaccaria*. Even though Winkler's paper was published earlier, it seems advisable to retain aplysin for the characterized compound **27** rather than for Winkler's toxic extract.

their following publication Irie and co-workers (1965) revealed the structure of a hydrocarbon laurene (**29**), which they had isolated from *L. glandulifera*.

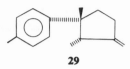

29

Laurene (**29**) was obtained by steam distillation of a methanolic extract of the algae as an oil, $[\alpha]_D^{23°}$ +48.7°. Its structure was secured by spectral analysis and by chemical degradation to the ketone **30**, the racemate of which was

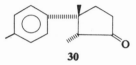

30

synthesized. In a footnote, the authors (Irie *et al.*, 1965b) make the interesting statement, "Laurene might be probably a precursor of aplysin, a sesquiterpene containing bromine isolated from the sea rat (*Aplysia kurodai* Baba) . . ., since *Sea rat* feeds on Laurencia." There is, in fact, a fair amount of ancillary evidence that indicates that the sea hares do not themselves elaborate the sesquiterpenoids but ingest them. However, proof for this hypothesis is at present lacking.

In their continued investigation of sesquiterpenoid constituents of the genus *Laurencia* Irie and collaborators (1966, 1970) isolated from *L. intermedia* three phenols—laurinterol, (**31a**), debromolaurinterol (**32a**), and isolaurinterol (**33a**)—as well as a trace of the debromo analog of isolaurinterol. Isolaurinterol (**33**) belongs to the structure type **25e**, whereas the other two compounds

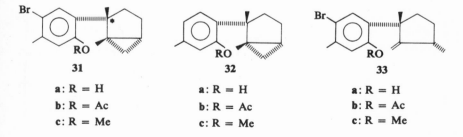

31	**32**	**33**
a: R = H	**a**: R = H	**a**: R = H
b: R = Ac	**b**: R = Ac	**b**: R = Ac
c: R = Me	**c**: R = Me	**c**: R = Me

are representatives of **25d**. Isolation and purification of the three compounds was best accomplished via their acetates: **31b**, mp 93°–93.5°, $[\alpha]_D$ +11.1°; and **32b**, oil, $[\alpha]_D$ −70°. Mild hydrolysis of the three acetates afforded the free phenols. The structures were secured by spectral analysis and by conversion of the three phenols to the corresponding aplysin derivatives. By means of *p*-toluenesulfonic acid in acetic acid, laurinterol (**31a**) and isolaurinterol (**33a**) were converted to aplysin (**27**); debromolaurinterol (**32a**) was transformed into debromoaplysin (**26**). Furthermore, lithium aluminum hydride effected the change from laurinterol (**31a**) into debromolaurinterol (**32a**), which in turn could be converted to laurinteryl acetate (**31b**) with bromine in acetic acid.

The acid-catalyzed rearrangement of laurinterol derivatives was studied in greater detail. Irie and co-workers (1969) noted that with *p*-toluenesulfonic acid in acetic anhydride rather than acetic acid debromolaurinteryl acetate (**32b**) furnished not debromoaplysin (**26**), but compound **34**, which, however, could be further transformed to debromoaplysin (**26**) with *p*-toluenesulfonic acid in acetic acid.

In a follow-up study of the acid-catalyzed rearrangements of laurinterol derivatives Suzuki *et al.* (1969) observed that laurinterol methyl ether (**31c**) when treated with *p*-toluenesulfonic acid in acetic acid led only to recovery of starting material. When the same compound (**31c**) was treated with 10% sulfuric acid in acetic acid, two racemized cyclopentene derivatives, **35** and **36**, resulted in a ratio of about 3.5:1. Corresponding cyclopentene derivatives were obtained when debromolaurinterol methyl ether (**32c**) was treated analogously. Under varied reaction conditions no compounds other than **35** and **36** were found, but prolonged reaction times favored the cyclopentene derivative represented by structure **35**.

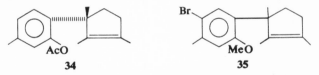

| 34 | 35 |

The stereochemical assignment of laurinterol as shown in **31** was determined by single-crystal X-ray diffraction of laurinteryl acetate, **31b** (Cameron *et al.*, 1967). In their full paper the crystallographers (Cameron *et al.*, 1969) further state that the boat conformation is preferred for laurinteryl acetate. The absolute configuration at the asterisked carbon in **31** has been related to the corresponding carbon in camphor (**37**) in the following manner. Irie *et al.* (1967) converted laurene (**29**) into (+)-cuparene (**38**), which had been shown by Enzell and Erdtman (1958) to have the same configuration at the starred

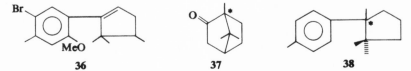

| 36 | 37 | 38 |

carbon as camphor (**37**), the absolute stereochemistry of which was proved to be as depicted (Allen and Rogers, 1966). Cuparene (**38**) represents the unrearranged farnesyl skeleton (structural type **25a**) and has not been found in either *Aplysia* or *Laurencia* or in any other marine source. It is, however, a heartwood constituent of various coniferous species in the plant family Cupressaceae (Enzell and Erdtman, 1958); its structural assignments have been confirmed by synthesis (Parker *et al.*, 1962).

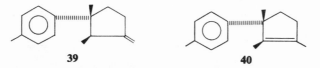

39   40

In their chemical conversion of laurene (**29**) into cuparene (**38**) Irie and co-workers (1967) used the natural ketone **30** as their key intermediate. Interestingly, they were unable to achieve the transformation back to laurene (**29**), but instead isolated a mixture of epilaurene (**39**) and isolaurene (**40**).

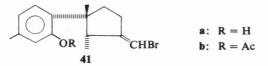

a: R = H
b: R = Ac

41

A variant of structure **39** is the sesquiterpenoid laurenisol (**41a**), which Irie and co-workers (1969a) isolated from *Laurencia nipponica*. Laurenisol (**41a**), $[\alpha]_D$ +85.9°, is unstable at room temperature, but forms a stable acetate (**41b**) mp 102.5°–103°, $[\alpha]_D$ +85.1°. When laurenisol (**41a**) was placed in a vacuum desiccator overnight, it was transformed into an oily mixture, from which two isomeric ethers, **42** and **43**, could be separated by silica gel chromatography in a ratio of 5:1. The rearrangement of ether **42** could be reversed, albeit with loss of bromine and acetylation of the phenol, by treatment of ether **42** with zinc and acetic acid, followed by acetylation. The resulting compound **44** is essentially a phenolic acetate of laurene (**29**).

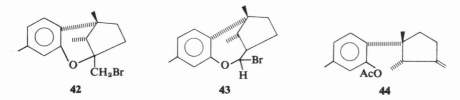

42   43   44

A close structural relationship of the sea hare constituents, the aplysins, and the sea weed constituents of the genus *Laurencia* has been amply demonstrated *in vitro* and has at least been suspected to exist *in vivo*. Indeed, Irie *et al.* (1969b) were able to demonstrate that in at least one *Laurencia* species, *L. okamurai*, aplysin (**27**), debromoaplysin (**26**), aplysinol (**28**), laurinterol (**31**), and debromolaurinterol (**32**) occur simultaneously. This interesting finding, however, does not exclude the possibility that sea hares are able to perform some synthetic steps toward the aplysins. A direct experiment to examine this point would be of interest because of its implications with respect to terpene modification and to mollusk metabolism. In addition to the five previously characterized compounds, *L. okamurai* yielded a compound with the unexpected composition $C_{14}H_{20}O_5Br_2$, mp 148.5°–149°, which— *mirabile dictu*—proved to be identical with johnstonol of $C_{15}H_{21}O_3Br_2Cl$ composition (Sims *et al.*, 1972).

In a recent paper on *Laurencia* constituents Irie's group (Suzuki *et al.*, 1970) isolated from *L. glandulifera*, which had previously (Irie *et al.*, 1965b) yielded laurene (**29**), the first example of an unrearranged farnesyl skeleton—however, one with a different cyclization pattern leading to structural type **25b**. The compound was named spirolaurenone, colorless oil, $[\alpha]_D$ −70.6°, and its structure was shown to be **45** on the basis of spectral—largely nmr—analysis and by chemical degradation.

Another spiro compound possessing the skeletal type **25c** has been isolated by Sims and co-workers (1971) from a Californian *Laurencia*, *L. pacifica*. This skeletal type, although new among marine sesquiterpenoids, has a

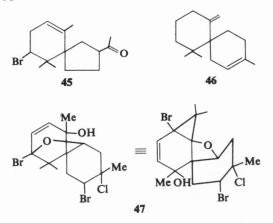

terrestrial representative in the sesquiterpene chamigrene (**46**), which Itô *et al.* (1967) had isolated from the essential oil of the leaves of a Formosan cypress tree, *Chamaecyparis taiwanensis*, and which was synthesized by Tanaka *et al.* (1967). The new algal sesquiterpenoid, pacifenol (**47**), differs from the known constituents of the Japanese *Laurencia* species not only by its skeleton, but by the fact that it contains covalent chlorine in addition to bromine. In fact, pacifenol (**47**) appears to be the first authentic chlorine containing organic compound to have been isolated from a marine source. This is indeed remarkable in view of the fact that the concentration of chlorine in seawater is nearly 300 times as great as that of bromine. The structure of pacifenol was secured by single-crystal X-ray techniques. The known compound laurinterol (**31a**), previously isolated by Irie *et al.* (1966, 1970) from *L. intermedia*, was also identified by Sims *et al.* (1971) as a minor constituent of *L. pacifica*.

TABLE 1.1

MARINE SESQUITERPENOIDS

Text no.	Name	mp (in degrees)	$[\alpha]_D°$	Reference
1	δ-Cadinol	137.5–138.5	−108	Irie *et al.* (1964)
2	γ₁-Cadenene	Oil	−19.6	Irie *et al.* (1964)
3	β-Bisabolene	Oil	+83.1	Weinheimer *et al.* (1968b)
4	Calamene	Oil	+55.4	Weinheimer *et al.* (1968b)
5	α-Muurolene	Oil	+67.1	Weinheimer *et al.* (1968b)
6	β-Elemene	Oil	+15.1	Weinheimer *et al.* (1970)
7	(−)-Germacrene-A	Oil	−3	Weinheimer *et al.* (1970)
8	(−)-β-Selinene	Oil	levo	Weinheimer *et al.* (1970)
9	9-Aristolene	Oil	+80.9	Weinheimer *et al.* (1968c)
10	1(10)-Aristolene	Oil	−78.5	Weinheimer *et al.* (1968c)
11	(+)-γ-Maaliene	Oil	+10.9	Weinheimer *et al.* (1968c)
12	(+)-α-Cubebene	Oil	+23.6	Weinheimer *et al.* (1968b)
13	(+)-β-Ylangene	Oil	9.2–10.6	Weinheimer *et al.* (1968b)
14	(−)-Copaene	Oil	−26.1	Irie *et al.* (1964)
15	Selinen-1β-ol (dictyopterol)	Oil	−30.8	Kurosawa *et al.* (1966)
16	β-Selinen-1-one (dictyopterone)	Oil	−12.5	Kurosawa *et al.* (1966)
17	β-Gorgonene	Oil	+13.9	Weinheimer *et al.* (1968c)
18	Furoventalene	Oil	—	Weinheimer and Washecheck (1969)
26	Debromoaplysin	Oil	−63	Yamamura and Hirata (1963), Irie *et al.* (1969b)
27	Aplysin	85–86	−85.4	Irie *et al.* (1969b)
28	Aplysinol	158–160	−55.6	Irie *et al.* (1969b)
29	Laurene	Oil	+48.7	Irie *et al.* (1965b)
31a	Laurinterol	54–55	+13.3	Irie *et al.* (1966, 1970)
32a	Debromolaurinterol	Oil	−12.2	Irie *et al.* (1966, 1970)
33a	Isolaurinterol	Oil	—	Irie *et al.* (1966, 1970)
41a	Laurenisol	Indefinite	+85.9	Irie *et al.* (1969a)
45	Spirolaurenone	Oil	−70.6	Suzuki *et al.* (1970)
47	Pacifenol	149–150.5	—	Sims *et al.* (1971)
48	Johnstonol	178	—	Fenical *et al.* (1971); Sims *et al.* (1972)

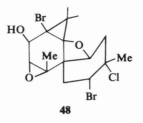

48

The same research group (Fenical *et al.*, 1971) reported the isolation of a related compound, johnstonol (**48**), from *L. johnstonii* collected in the Gulf of California. The structure of johnstonol was established by single-crystal X-ray techniques (Sims *et al.*, 1972).

Table 1.1 on page 17 contains a list of characterized marine sesquiterpenoids.

B. Diterpenoids

Only four examples of true diterpenoid compounds have so far been reported from marine organisms, three from gorgonians and one from a sea hare.

1. THE C$_{20}$ COMPOUNDS

The sea hare constituent, designated aplysin-20, is a bicyclic diterpenoid, which was isolated from *Aplysia kurodai* along with the three sesquiterpenoids aplysin (**27**), debromoaplysin (**26**), and aplysinol (**28**). The structure of aplysin-20, mp 146°–147°, $[\alpha]_D^{15°} -78.1°$, was determined by X-ray diffraction techniques (Matsuda *et al.*, 1967) to be **49**. Aplysin-20 possesses a normal (unrearranged) isoprenoid skeleton; but as the authors point out, the compound has the unusual features of an axial hydroxyl at C-8 and an equatorial bromine at C-3. The full paper, which includes all spectral details, was published more recently (Yamamura and Hirata, 1971). In this paper the authors mention that they isolated aplysin-20 from *A. kurodai* collected at Hokkaido, but not at Sugashima, an observation that may be related to the nature or seasonality of the mollusk's diet.

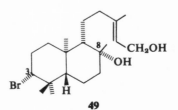

49

Perhaps the first mention in the literature of a marine-derived diterpenoid—although unrecognized as such at the time—may be found in a paper of the Oklahoma group (Ciereszko *et al.*, 1960). During pentane or ether extraction of the gorgonian *Plexaura crassa* a substance crystallized out, mp 144°–144.5°, $[\alpha]_D^{25°}$ +70.7°, which was called crassin acetate. A summary of the chemical work (Weinheimer *et al.*, 1968) and the details of the X-ray determination of crassin *p*-iodobenzoate (Houssain and van der Helm, 1969) have been published. The structure of crassin acetate (**50**), as are the structures of the other

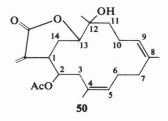

50

two diterpenoids isolated from gorgonians, are derivatives of the fourteen-membered carbocyclic thunbergane (syn. cembrane), a structure type that has only recently been encountered in the oleoresin of Douglas fir (Erdtman *et al.*, 1968; Kimland and Norin, 1968). Weinheimer *et al.* (1968b) isolated crassin acetate from three additional gorgonians, *Pseudoplexaura porosa, P. wagenaari,* and *P. flagellosa,* mp 138°–140°, $[\alpha]_D^{25°}$ +70.4°.

The second of the cyclotetradecane-based diterpenoids from a gorgonian is eunicin, mp 154°–155.5°, $[\alpha]_D^{25°}$ −89.4° (Weinheimer *et al.*, 1968a). It was isolated from *Eunicea mammosa* collected in Bimini. Its structure **51a** was elucidated by spectral and chemical techniques (Weinheimer *et al.*, 1968)

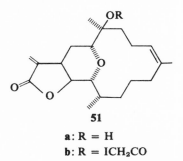

51

a: R = H
b: R = ICH₂CO

as well as by X-ray diffraction (Houssain *et al.*, 1968) of the crystalline iodo-acetate (**51b**), mp 149°–150°. Remarkably, eunicin occurs in the animal to the extent of 1% of dry weight.

The third compound in the group, jeunicin, with an mp of 139°–141°, $[\alpha]_D^{25°}$ +12.8°, was also isolated from *Eunicea mammosa*, but from specimens

collected at Jamaica. A tentative structure, **52**, has been suggested by Weinheimer *et al.* (1968).

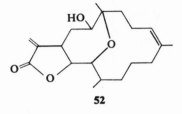

52

2. The C_{19} Compounds

An interesting pair of C_{19} hydrocarbons, apparently degraded diterpenoids,
have long been known as constituents of the liver oils of sharks and whales.
The more widely distributed of the two, pristane, was first discovered by
Tsujimoto (1917) in the liver oil of the basking shark *Selache maxima* (syn.
Cetorhinus maximus). Not surprisingly, it was incompletely characterized
at the time. Some thirty years later Sørensen and Mehlum (1948) reisolated
pristane from the liver oil of the basking shark. Cryoscopic molecular weight
determinations pointed to a $C_{19}H_{40}$ composition and its extremely low
melting point (below dry ice temperature) indicated a highly branched
structure. The most likely structure, norphytane (2,6,10,14-tetramethylpenta-
decane), **53**, was synthesized from the ubiquitous plant alcohol phytol (**54**)
and was shown to be identical with pristane (Sørensen and Sørensen, 1949).

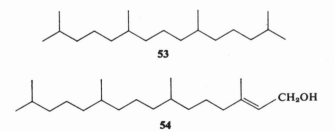

53

54

The second hydrocarbon, the olefin zamene, was first partially characterized by Tsujimoto (1935) and shown to be 2,6,10,14-tetramethyl-1-penta-
decene (**55**) by degradation to formaldehyde and the known 6,10,14-trimethyl-
2-pentadecanone (**56**) (Christensen and Sørensen, 1951).

Blumer *et al.* (1963) isolated pristane in high concentrations from three
copepods of the genus *Calanus*. It appears likely that these planktonic
crustaceans are the immediate source of pristane in sharks and whales.

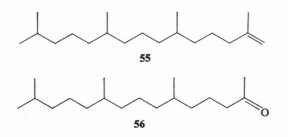

3. THE C_{21} COMPOUNDS

Two related terpenoids, each possessing twenty-one carbon atoms, have recently been isolated by Fattorusso and co-workers (1971) from the Mediterranean sponge *Spongia nitens*. The major constituent, a colorless oil of composition $C_{21}H_{24}O_4$, was named nitenin and was shown to have structure **57**. The conventional head-to-tail isoprenoid construction with one additional carbon atom (C-21) is indicated by the heavy lines. Structure **57** of nitenin was unambiguously established by spectral, particularly nmr, data and by ozonolysis of niteninic acid that had been prepared from nitenin by opening of the lactone ring in alkaline solution. The chirality of the sole asymmetric carbon atom C-11 was shown to be *R* by mild lithium aluminum hydride reduction of nitenin (**57**) to the diol **58**, followed by partial asymmetric esterification as developed by Horeau (1961, 1962).

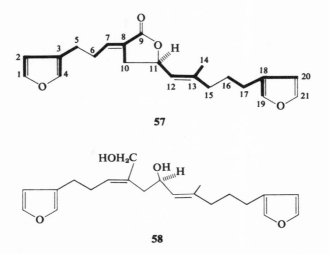

The second, minor, constituent was named dihydronitenin. Its structure (**59**) was deduced on spectral grounds and was confirmed by sodium borohydride reduction of nitenin (**57**).

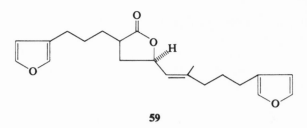

59

The authors (Fattorusso *et al.*, 1971) speculate on the biogenetic origin of these interesting C_{21} compounds. They prefer the idea that nitenin and dihydronitenin arose from degradation of a larger terpenoid over the possibility of synthesis by addition of a C-1 unit, since the Italian workers now isolated several C_{25} (sester) terpenoids from the sponges, *Ircinia oros* (Cimino *et al.*, 1972) and *I. fasciculata* (Cafieri *et al.*, 1972).

Table 1.2 lists the marine diterpenoids and the related C_{19} and C_{21} compounds.

TABLE 1.2

MARINE DITERPENOIDS

Text no.	Name	mp (in degrees)	$[\alpha]_D°$	Reference
49	Aplysin-20	146–147	−78.1	Matsuda *et al.* (1967)
50	Crassin acetate	144–144.5	+70.7	Ciereszko *et al.* (1960)
51	Eunicin	154–155.5	−89.4	Weinheimer *et al.* (1968a)
52	Jeunicin	139–141	+12.8	Weinheimer *et al.* (1968b)
53	2,6,10,14-Tetramethyl-pentadecane (pristane)	—		Sørensen and Sørensen (1949)
55	2,6,10,14-tetramethyl-1-pentadecene (zamene)	—		Christensen and Sørensen (1951)
57	Nitenin	—	−45.4	Fattorusso *et al.* (1971)
59	Dihydronitenin	—	−25.2	Fattorusso *et al.* (1971)

C. Triterpenoids

All except two of the triterpenoid compounds that have so far been isolated and characterized from marine sources are elaborated by members of a single phylum of exclusively marine invertebrates, the echinoderms (Echinodermata). The phylum comprises about 5300 described species and includes some of the most familiar representatives of sea life, e.g., sea stars (starfish) and sea urchins. The phylum is characterized by radial symmetry, and some of the animals are spectacularly beautiful in form and color.

The trivial names of the five classes in the phylum are descriptive of their appearance and thereby indicative of the divergence among the classes. No wonder that a lively controversy exists among marine zoologists with respect to phylogenetic relationships in the phylum. All chemical indicators that have so far been studied seem to point to a close relationship between sea urchins (Echinoidea) and brittlestars (Ophiuroidea), and similarly between sea stars [popularly called starfish, (Asteroidea)] and sea cucumbers (Holothuroidea). The fifth class, the feather stars or sea lilies (Crinoidea), appears to be unique. This purely chemical view of crinoid phylogeny receives some support from the fact that the crinoids are the most ancient—and in some ways the most primitive—of the living echinoderms.

1. HOLOTHURINS

The sea cucumbers or holothurians are a class of some 500 species. To the human eye they are perhaps the least attractive of the echinoderms in shape and color, although the dried body wall of certain large species (e.g., *Thelenota ananas, Stichopus variegatus, Holothuria atra*) is an East Asian (mostly Chinese) culinary delicacy traded under the name *trepang* or *bêche-de-mer*. They are further distinguished from the other echinoderms by their general lack of calcareous body armor. Nevertheless, sea cucumbers possess some unique features that make them attractive, at least to scientists. One of these features is a well-documented symbiotic relationship between the pearl fish (*Carapus* spp.) and certain sea cucumbers, which play host to the fish in their cloaca. The fish leaves the host at night while it searches for food, then returns to its holothurian shelter. Another unusual feature, of greater immediate chemical interest, is a defensive mechanism associated with the so-called Cuvierian tubules or glands. Some species of holothurian in the genera *Holothuria* and *Actinopyga* possess these tubules, which are attached to the base of the respiratory system. These sticky tubules are shot out of the sea cucumber's anus toward a predator who finds himself entangled and disabled in a sticky mesh of adhesive threads. The tubules will regenerate after discharge.

According to Halstead (1965) the earliest record of the toxic nature of the Cuvierian tubules date back to 1880, but modern research into the constituents of sea cucumbers was probably started in 1929 by Yamanouchi* (1955).

Apparently one of Yamanouchi's earliest observations was related to the ability of aqueous extracts of *Holothuria vagabunda* to kill fish when placed

---

* Yamanouchi's earlier publications on his research (1929, 1942, 1943) are cited by him, and occasionally by others. Some are in Japanese and are published in periodicals that are not readily available in United States libraries.

in aquaria. In a subsequent paper,* Yamanouchi (1942) surveyed eight species of sea cucumber and noted that all eight, including the one that is traditionally eaten, contain in their body walls a toxic principle in varying amount. From the most toxic species, *Holothuria vagabunda*, he isolated a crystalline toxin which he named holothurin 1a. Yamanouchi (1942) further noted its water solubility, its hemolytic property, and its similarity to vegetable saponins.

Nigrelli (1952), undoubtedly unaware of Yamanouchi's prior publications, also proposed the name holothurin for the active principle in the Cuvierian tubules of *Actinopyga agassizi*. Nigrelli and Zahl (1952) noted the antitumor activity of holothurin and its ability to kill fish, even members of the hardy genus *Carapus*, the sea cucumber's frequent parasitic lodger.

Pioneer work in further definition of holothurin and in structural elucidation of the individual constituents was carried on by Nigrelli's group at the New York Aquarium in conjunction with a molecular structure group at Mount Sinai Hospital, New York, notably Chanley and Sobotka. By 1955 (Nigrelli *et al.*, 1955; Chanley *et al.*, 1955) it had become apparent that holothurin was a steroidal (more accurately, triterpenoid) glycoside which was hydrolyzable into closely related aglycones, several monosaccharides, and sulfuric acid. While holothurin was transparent in the ultraviolet, acid hydrolysis generated in the aglycone a chromophore with λ_{max} 244 nm, suggestive of a heteroannular diene. Nigrelli and Jakowska (1960) have concisely summarized much of the early work. Matsuno and Yamanouchi (1961) in a follow-up of Yamanouchi's earlier observations (1942, 1955) described the hydrolysis of the holothurin from *Holothuria vagabunda*. It led to the isolation of a genin called holothurigenin, mp 301°, $[\alpha]_D$ +14.9°, and composition $C_{30}H_{44}O_5$. Matsuno and Yamanouchi (1961) recognized the genin as a triterpenoid possessing three hydroxyl, a lactone, and a heteroannular diene function.

Chanley and co-workers (1959) took advantage of the property of holothurin—which it shares with digitonin and other vegetable saponins—to form a complex with cholesterol and precipitated about 40% of the original glycosidic mixture as a cholesterol complex. Treatment with pyridine recovered the glycosides, which were termed holothurin A. Upon acid hydrolysis this neutral, water-soluble, nonreducing mixture was transformed into a mixture of at least four water-insoluble aglycones, sulfuric acid, and water-soluble reducing sugars. The sugars were identified as D-xylose (**60**), D-glucose (**61a**), 3-methoxy-D-glucose (**61b**), and D-quinovose (6-deoxy-D-glucose) (**62**). Enzymatic hydrolysis at pH 5.2 with an extract of *Helix pomatia* (a mollusk) over 189 hours established the molecular sequence as

* I am grateful to Mr. Y. Kato for securing and translating this paper.

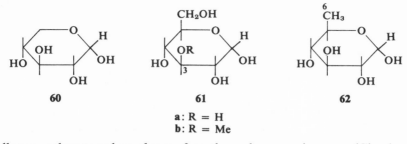

60

61

a: R = H
b: R = Me

62

follows: aglycone-xylose-glucose-3-methoxyglucose-quinovose (Chanley *et al.*, 1960). The authors deduced on the basis of circumstantial evidence that the sulfuric acid moiety was attached to the aglycone rather than to a sugar. Subsequently this point turned out to be incorrect. Chanley and Rossi (1969b) mention in passing that the xylose molecule is the locus of attachment for the sulfuric acid.

The molecular structures of two of the aglycones, the first marine triterpenoids to be elucidated, were disclosed by Chanley and co-workers (1966). Isolation of the glycosides was no longer effected via the cholesterol complex, but simply by precipitation with a saturated hydrocarbon solvent from a pyridine solution of carefully dried powdered Cuvier glands. Resulting crude holothurin constituted ca. 70% by weight of the glands. Holothurin was hydrolyzed by hot 3 *N* hydrochloric acid from which the mixture of aglycones, called holothurinogenins, precipitated. Further purification yielded ca. 30% of the weight of crude holothurin as holothurinogenins. Chromatographic separation of the holothurinogenins furnished two pure compounds, 22,25-oxiodoholothurinogenin (**63a**) in about 20% of the aglycone mixture,

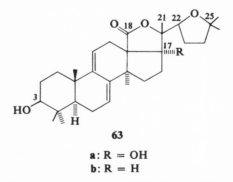

63

a: R = OH
b: R = H

17-deoxy-22,25-oxidoholothurinogenin (**63b**) (about 10%), and a mixture of at least four additional aglycones that appear to differ from **63** only in the side chain. This point was ascertained by lithium aluminum hydride reduction of the mixture followed by lead tetraacetate oxidation, and by characterization of the resulting nonvolatile (ring system) and volatile (side chain) ketones.

The structure of **63a**, 22,25-oxidoholothurinogenin, $C_{30}H_{44}O_5$, was carefully and convincingly deduced by Chanley *et al.* (1966) on the basis of chemical degradation and spectral determinations.

Presence of a trans heteroannular diene system in the genin, but—as was pointed out earlier—absent in the glycosides, was based primarily on the characteristic triple band in the ultraviolet: a maximum at 243 nm (ε14400) and two prominent shoulders at 237 nm (ε13430) and 252 nm (ε10390). Triple substitution of each of the two double bonds was inferred from the presence in the nmr spectrum of the 3-acetate of only two broad one-proton signals at δ5.25 and δ5.51.

Two of the five oxygen functions in **63a** are represented by hydroxyl groups—one secondary that can be oxidized to a ketone, and one tertiary. 22,25-Oxidoholothurinogenin forms monoesters that retain hydroxyl absorption in the infrared (ca. 3565 cm^{-1}); under forcing conditions it forms a bistrifluoroacetate, which is devoid of hydroxyl absorption. A five-membered lactone accounts for two more oxygen atoms. Its major support rests on the infrared band at 1763 cm^{-1}. The lactone appeared to be hindered since the aglycone was recovered unchanged after prolonged refluxing with alcoholic potassium hydroxide. The ethereal nature of the fifth oxygen atom was inferred by exclusion and was positively indicated by a broad one-proton triplet nmr signal centered at δ4.23 ($J = 6$ Hz), which was assigned to a methine hydrogen flanked by a cyclical ether oxygen and a methylene group.

Seven quarternary methyl groups could be distinguished in the nmr spectrum of **63a**. Since a "normal" triterpenoid compound has eight such groups, it was a reasonable assumption that one methyl group had been oxidized and had become part of the lactone. Presence of two olefinic linkages, a lactone and an ether ring pointed to a tetra- rather than a penta-cyclic triterpenoid. Tetracyclic triterpenoids of the dammarane or isoeuphane types were excluded since neither will accommodate a trans heteroannular diene bearing only two vinyl hydrogen atoms. The methyl signals ranged from 0.91 to 1.38 ppm in deuteriochloroform and from 0.90 to 1.59 ppm in pyridine. The signal farthest downfield was assigned to the C-21 methyl group that is attached to C-20, which in turn is bonded to the lactonic oxygen atom. The γ-lactone ring would then include the oxidized C-18 methyl. This assumption agrees with the hindered nature of the lactone as well as with the fact that an unoxidized C-18 methyl would give rise to an nmr signal farther upfield than 0.90 ppm. Two downfield methyl signals at 1.26 and 1.78 ppm (in $CDCl_3$) that were incompletely resolved were compatible with the *gem*-dimethyl group at C-25 that would be linked to the ether oxygen in line with their chemical shifts. The other carbon (C-22) that is linked to the ether oxygen atom bears a single hydrogen atom and has already been assigned. Excellent support for this unusual side chain came from the mass spectrum of **63a**, the strong peak of which at m/e 99 corresponding to $C_6H_{11}O$ was assigned to fragment **64**.

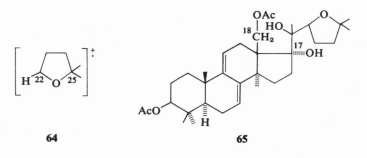

64   65

Strong chemical evidence for the nature and position of the lactone was derived from lithium aluminum hydride reduction of **63a** to a tetrol and its conversion to a diol diacetate (**65**), the structure of which is in accord with its spectral properties. Specifically, two new one-proton doublets centered at 3.86 and 4.30 ppm ($J = 12$ Hz) are readily assigned to two nonequivalent hydrogen atoms at C-18.

The diol diacetate **65** was further degraded by lead tetraacetate oxidation to two ketones without loss of carbon: a 22-carbon compound **66** and a volatile

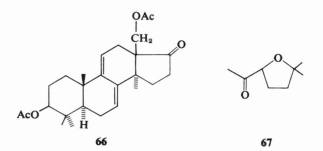

66   67

fragment (**67**) having eight carbon atoms. The spectral properties of compound **66** demonstrated conclusively that the side chain in the holothurinogenin **63** is attached to C-17 as is the tertiary hydroxyl.

The 7:8, 9:11 diene system was confirmed by several chemical transformations. Perhaps the most direct method was by oxidation of the acetate of **63a** with chromic oxide to a yellow trans ene-dione **68**, λ_{max} 280 nm (ε6390). This result is characteristic of tetracyclic triterpenoids possessing a 7:8, 9:11 diene system (Ourisson *et al.*, 1964).

Spectral comparison with other triterpenoid derivatives in the lanosterol and agnosterol series established the typical 3β-hydroxy-4,4-*gem*-dimethyl system for **63a**.

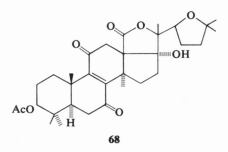

68

The structure of the second constituent, 17-deoxy-22,25-holothurinogenin (**63b**), was readily deduced (Chanley *et al.*, 1966) by preparation of several derivatives and by spectral comparisons.

A third holothurinogenin, griseogenin (**69**), was isolated in addition to the known 22,25-oxido compound (**63a**) by Tursch *et al.* (1967) from the Cuvier

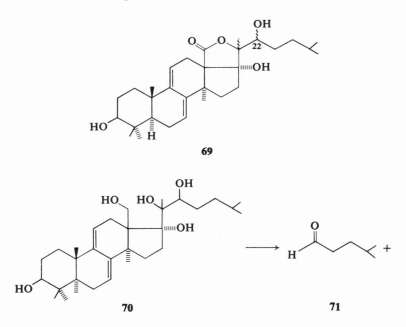

glands and the body wall of the sea cucumber *Halodeima grisea*. The key degradation that led to structure **69** involved lithium aluminum hydride reduction of the mono- or diacetate to a pentaol (**70**), which was cleaved by periodic acid in aqueous methanol. The resulting volatile aldehyde was identified as 4-methylpentanal (**71**). The nonvolatile ketoacetate **72a** was compared with authentic **72b** upon acetylation. Its generation from **70** proceeds through the following plausible sequence.

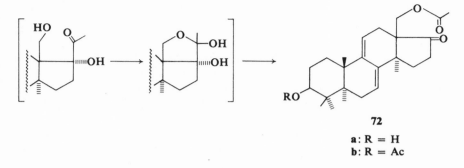

72

a: R = H
b: R = Ac

A number of closely related aglycones have since been isolated from several species of holothurians. Roller *et al.* (1969) reported three new genins from the sea cucumber *Bohadschia koellikeri*—seychellogenin (**73a**), koellikerigenin (**73b**), and ternaygenin (**73c**). The three compounds could be

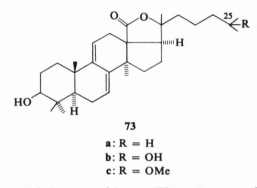

73

a: R = H
b: R = OH
c: R = OMe

readily interconverted, **b → a** and **b → c**. The great ease of methylation of koellikerigenin (**73b**) to ternaygenin (**73c**) suggested to the authors (Roller *et al.*, 1969) that ternaygenin might be an artifact because the conditions under which the C-25 hydroxy group is methylated are identical with those under which the natural glycosides are hydrolyzed.

A fourth holothurinogenin, praslinogenin (**74**), was isolated by Tursch *et al.* (1970) from the same animal, *Bohadschia koellikeri*. A comparison of

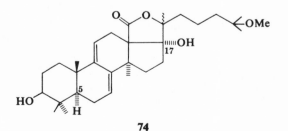

74

spectral data showed praslinogenin (**74**) to be a hydroxylated ternaygenin (**73c**). That praslinogenin (**74**) was indeed the 17-hydroxy compound could be demonstrated by comparing the chemical shift of one of the terminal methyl groups—it moves downfield from 1.00 to 1.15 ppm when one goes from ternaygenin (**73c**) to praslinogenin (**74**). The same value (1.15 ppm) is observed for a terminal methyl group in other 17α-hydroxy compounds, e.g., griseogenin (**69**). Again, it is likely that praslinogenin (**74**) is an artifact produced during glycoside hydrolysis.

Habermehl and Volkwein (1968) examined several species of Mediterranean sea cucumbers and succeeded in isolating 22,25-oxidoholothurinogenin (**63a**) as well as its 17-deoxy analog (**63b**) from *Holothuria tubulosa*. Paper chromatography of the sugars indicated the presence of only an aldopentose and the absence of sulfuric acid.

From another Mediterranean species, *H. polii*, Habermehl and Volkwein (1970) were able to separate and identify by spectral comparison six aglycones after acid hydrolysis. In addition to the known compounds 22-25-oxidoholothurinogenin (**63a**) and its 17-deoxy analog (**63b**), griseogenin (**69**), praslinogenin (**74**), and its deoxy analog ternaygenin (**73c**) the German workers demonstrated the presence of the hypothetical parent compound of the

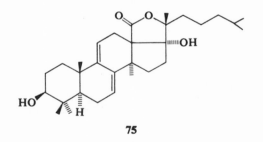

75

group, holothurinogenin (**75**). Interestingly, the aglycones of *H. polii* include the two 25-methoxy compounds (**73c, 74**) without the corresponding 25-hydroxy "precursors" in spite of the fact that aqueous hydrochloric acid is the hydrolytic agent in an operation that is separate from the original aqueous methanolic extraction. This would indicate that 25-methoxy compounds need not be artifacts. Habermehl and Volkwein (1970) assign β-configuration (i.e., trans to the 17α-hydroxy group) to the C-21 methyl group, since this methyl group exhibits the same chemical shift in the 22,25-oxido compound (**63a**) and in the 25-methoxy compound (**74**).

Habermehl and Volkwein (1971) examined a third Mediterranean sea cucumber, *Holothuria forskåli*, and isolated from the Cuvierian organs three known genins—22,25-oxidoholothurinogenin (**63a**), 17-deoxy-22,25-oxidoholothurinogenin (**63b**), and 25-methoxyholothurinogenin (praslinogenin) (**74**)—albeit in very low yield.

The same authors (Habermehl and Volkwein, 1971) make the sensible suggestion to discontinue the confusing and no longer needed trivial names in this series of compounds and derive all names by standard nomenclature from the parent compound holostanol (**76**), which in turn is identical with $3\beta,20\alpha_F$-dihydroxy-5α-lanostan-18-carboxylic acid-(18 → 20)-lactone.

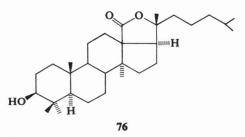

76

The structures of all eight holothurinogenins had been interrelated and/or directly compared; all experimental evidence was consistent with the structural assignments; and Chanley *et al.* (1966) had ascertained the key features of 22,25-oxidoholothurinogenin (**63a**) and its deoxy analog (**63b**) by multiple experimental routes. Yet a direct interrelation of this new group of triterpenoids with a lanostane derivative was highly desirable because the holothurinogenins were assumed to be variants of the lanosterol structural type. This interrelation was performed by Roller *et al.* (1969, 1970). The simplest known

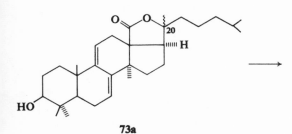

73a

$\longrightarrow$

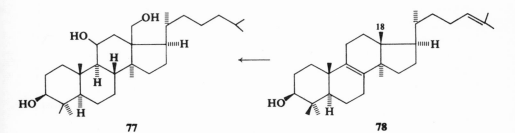

77   **78**

holothurinogenin was seychellogenin (**73a**). It was chosen for degradation to an intermediate **77** that was also prepared from lansterol (**78**). The functionalization of C-18 of lansterol (**78**) was achieved through a photochemical reaction with lead tetraacetate and iodine. This successful interconversion confirmed all structural features of seychellogenin (**73a**) and by extension of all other known holothurinogenins with exception of the configuration at C-20. Stereochemistry at C-20 was still in doubt since the degradation of **73a** destroyed the chirality when the ring-opened triol was dehydrated. Hydrogenation of the $\Delta^{20,21}$-olefin led to a mixture of epimers which was subsequently separated, and the epimer which had the lanosterol stereochemistry at C-20 was utilized.

The inconsistency of the ultraviolet spectral properties of the natural glycosides (no absorption beyond 212 nm) when compared with the holothurinogenins (triple band centered at 244 nm characteristic of a heteroannular diene) had made it clear at the outset of these investigations that the structure of the holothurinogenins could not be the structure of the intact triterpenoid glycosides prior to acid hydrolysis. Obviously the heteroannular diene system was generated in the course of the acid treatment, but the lactone function and the 22,25-oxide bridge were also suspect as potential artifacts that had arisen during hydrolysis. Chanley and Rossi (1969a,b) resolved these points of uncertainty in the following manner. When holothurin A—the crude glycosidic mixture—was hydrolyzed in 0.2 N hydrochloric acid in methanol at 50°, hydrolysis was complete after 76 hours without significant generation of the diene chromophore. The resulting aglycones were a mixture of mono- and dimethoxy *neo*-holothurinogenins, which by the conventional strong acid treatment could be converted to the known holothurinogenins.

Holothurin A, the mixture of natural glycosides, contains only one methoxy group as 3-methoxy-D-glucose. The methoxy group(s) of the *neo*-holothurinogenins must therefore have arisen as a result of methanolysis. By careful spectral analysis the structures of the *neo* compounds were shown to be **79a**, **79b**, **80**, **81**, and the dimethoxy compound **82**. Compound **80** represented the precursor of griseogenin (**69**), whereas compound **81** exhibited a new side chain. Compound **81** on strong acid treatment was converted to the corresponding $\Delta^{24,25}$-holothurinogenin, which also was the compound into which the dimethoxy-*neo*-holothurinogenin **82** was transformed. Chanley and Rossi (1969a) further demonstrated that the 22,25-oxido side chain in compounds **63a** and **b** was a naturally occurring structural feature and that the 25-methoxy group of **82** was generated from a precursor with a 24,25-double bond (**81**).

The 12-methoxy group that was common to all *neo* compounds most likely resulted from an allylic 12-hydroxy group. This supposition was con-

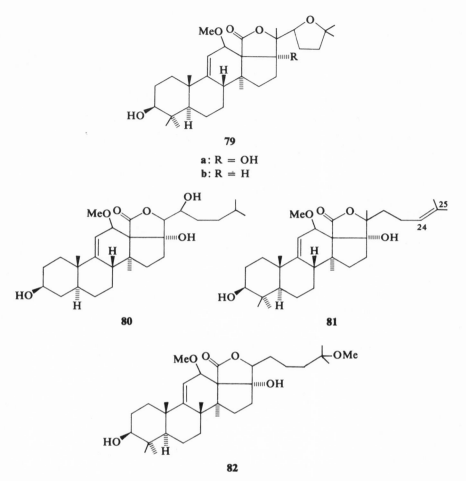

79

a: R = OH
b: R = H

80

81

82

firmed by carrying out the hydrolysis in dioxane rather than methanol. Nmr studies revealed the configuration of the 12-methoxy group to be *beta*. On the basis of arguments based on nmr solvent shifts Chanley and Rossi (1969a) were further led to conclude that the configuration of the C-21 methyl group was analogous to lanosterol, i.e., behind the plane of the lactone ring.

The great ease with which the 12-hydroxy group in holothurin A is converted to the 12-methoxy compounds suggested to Chanley and Rossi (1969b) that the configuration of the original hydroxy group might be axial, i.e., α, and that the β-methoxy disposition of the *neo* series of compounds is a result of epimerization during acid treatment. Chanley and Rossi (1969b) proved this point by enzymatic hydrolysis of desulfated holothurin A (holothurin A itself proved to be unsuitable), which led to some aglycones with the 12α-hydroxy configuration.

It is therefore evident that the 7:8, 9:11 diene system of the holothurino-genins is generated from a 9,11-en-12α-ol precursor, while the lactone and the sometimes tetrahydrofuran functions are present in the natural compounds.

The presence of a holothurin, designated holothurin B, that possesses the same mixture of aglycones as does holothurin A, but differs from it in its sugars was established by Yasumoto *et al.* (1967). These workers isolated from *Holothuria vagabunda* and *H. lubrica* holothurins A and B. Holothurin B, on acid hydrolysis, furnished only two sugars, D-xylose (**60**) and D-quino-vose (**62**), in addition to sodium hydrogen sulfate.

A likely variant of the holothurins is Shimada's (1969) holotoxin, which he isolated from the body wall of the sea cucumber *Stichopus japonicus*. This colorless crystalline substance shows high activity against pathogenic fungi, a property that distinguishes it from the holothurins. The reported physical and chemical properties, however, would tend to place holotoxin in a close structural relationship with the holothurins.

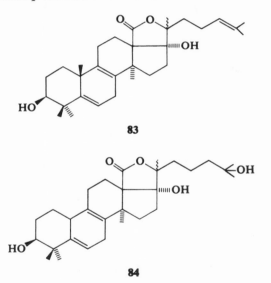

83

84

Furthermore, it is possible that Shimada's (1969) holotoxin is identical with, or closely related to, one of the glycosides that Elyakov *et al.* (1969) isolated from the same holothurian, *Stychopus japonicus*.* The Russian workers report the isolation of two glycosides, stichoposide A and C. Hydroly-sis of stichoposide A with 12% aqueous hydrochloric acid furnished two aglycones—stichopogenin A_2 and A_4. Elyakov *et al.* (1969) assign structure **83** to stichopogenin A_2 and structure **84** to stichopogenin A_4, essentially on the basis of spectral, largely nmr, data. No direct comparison of these compounds

* Presumably *Stychopus* and *Stichopus* represent only one genus.

with Shimada's (1969) holotoxin has been made, but the nonconjugated diene system in **82** and **84** is compatible with the lack of UV absorption that Shimada (1969) reports for holotoxin.

In an interesting attempt to discover the biological precursors of the holothurins (and the holothurian sterols as well) Nomura *et al.* (1969a,b) recently isolated from *Stichopus japonicus* and from *Holothuria tubulosa* the triterpenoids lanosterol (**78**) and cycloartenol (**85**) as their epoxide propionates. In an incorporation experiment on *S. japonicus*, acetate that was labeled at C-1 and C-2 with carbon-14 was injected into the body cavity of 40 holothurians. At the end of 11 days 0.53% of the radiocarbon was present in the unsaponifiable lipids, but neither lanosterol (**78**) nor cycloartenol (**85**) carried a label; squalene (**86**), however, did carry radioactivity. Nomura *et al.* conclude from this experiment that sea cucumbers do not synthesize the holothurins, but ingest them with their food.

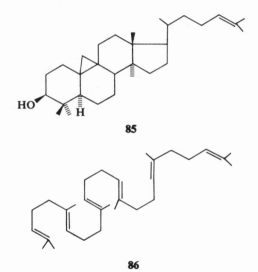

85

86

It is worth noting that squalene (**86**), the biogenetic precursor of triterpenoids and steroids, was first isolated in 1916 (Tsujimoti, 1916) from the liver oils of several species of shark. It has since been found to occur widely, as, e.g., in many seed oils of flowering plants and in brewer's yeast. Karrer and co-workers (1930) established its structure and synthesized it (Karrer and Helfenstein, 1931).

2. THE ASTEROSAPONINS

The echinoderm class of the asteroids (sea stars, starfish) is quite different in appearance from the holothurians. Frequently the sea stars, which possess a calcareous shell or test, are very beautiful in shape and color. Commonly they have five arms, although the most notorious species of recent years, *Acanthaster planci*, the coral-eating one, does not. Few if any casual observers would suspect that asteroids and holothurians have many common characteristics.

Hashimoto and Yasumoto (1960) first recognized the occurrence of a "saponin" in the asteroids. Several years later Hashimoto and his group (Yasumoto *et al.*, 1966) surveyed all five classes of echinoderms and found that only the holothurians and the asteroids contain these compounds, while the echinoids, ophiuroids, and crinoids do not. This finding reinforces the suspected close relationship of sea cucumbers and sea stars, which has also been suggested on the basis of two chemical parameters: Δ^7-sterols by Gupta and Scheuer (1968) and naphthoquinone pigments by Singh *et al.* (1967).

Following their recognition of sea star (astero) saponins Hashimoto and his group studied a number of them. From *Asterias amurensis* Yasumoto and Hashimoto (1965) isolated a mixture of six triterpenoid glycosides and designated the major component asterosaponin A. On acid hydrolysis the glycoside yielded 2 moles of D-quinovose (**62**) and 2 moles of D-fucose (**87**) in addition to sulfuric acid. The aglycone presented spectral evidence for a heteroannular diene, a characteristic structural feature of the holothurinogenins, but lacked spectral evidence for a lactone. A second glycoside, asterosaponin B, was isolated by Yasumoto and Hashimoto (1967) from the same asteroid. Acid hydrolysis furnished 2 moles of D-quinovose (**62**), 1 mole each of D-fucose (**87**), D-xylose (**60**), and D-galactose (**88**), and sulfuric acid. Prior to hydrolysis asterosaponin B exhibits an ultraviolet band at 248 nm and infrared bands at 1700 and 1640 cm^{-1}. These data indicate that asterosaponin B differs distinctly from the known holothurins and from asterosaponin A. Two aglycones have been partially characterized; one contains a conjugated, the other an isolated carbonyl group.

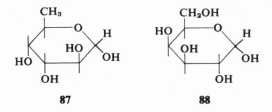

Friess and co-workers (1968) have recently made physiological comparisons of the characterized holothurins A and B and the asterosaponins A and B. Their experiments led them to conclude that the key to differential physiological activity lies in the nature of the sugars, their sequences, and the position in the chain of the sodium hydrogen sulfate. This finding no doubt has bearing on the reported unusual physiological properties of holotoxin, as pointed out by Shimada (1969).

The first two asterosaponins have now been fully characterized (Turner *et al.*, 1971) and surprisingly turn out to be not triterpenoidal but steroidal glycosides. These compounds will therefore be discussed in Chapter 2.

3. KNOWN TRITERPENOIDS

a. Friedelin

So far only two triterpenoids that are known constituents of terrestrial plants have been isolated from marine sources. The first such compound, friedelin (**89**) was identified by Tsuda and Sakai (1960) after its isolation

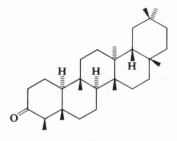

89

from a blue-green alga *Monostroma nitidum*. Although it was first isolated from cork (Friedel, 1892) in the last century, its structure was not fully elucidated until 1955 (Corey and Ursprung, 1955a,b; Dutler *et al.*, 1955).

b. Taraxerol

Another triterpenoid, taraxerol (**90**), which is widely known from terrestrial sources, was recently isolated from the green alga *Caulerpa lamourouxii* by Santos and Doty (1971). Its constitution was deduced by Beaton and co-workers (1954), who also partially synthesized it from β-amyrin (**91**) (Beaton *et al.*, 1955a,b).

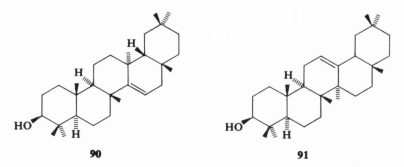

90   **91**

4. A NORTRITERPENOID

Although the group of antibiotics known as the cephalosporins are sometimes listed among marine natural products (e.g., Baslow, 1969) it is question-
able whether these compounds are in fact of marine origin. Such a compound,
a nortriterpenoid of somewhat dubious marine ancestry, is cephalosporin
P_1 (**92**). It will be briefly mentioned here, but will be omitted from Table 1.3,

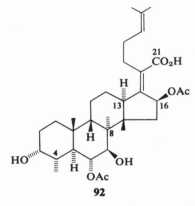

92

which lists the marine triterpenoids. According to Crawford *et al.* (1952),
who first cultured the organism that produces it, the antibiotic was first
isolated by Brotzu in 1948 from a microorganism, *Cephalosporium* sp.,
perhaps *C. acremonium*, that was collected near a sewage outfall off the coast
of Sardinia. Early structural work (Burton *et al.*, 1956) indicated a steroidal
character of C_{32} composition. Mass spectral (Halsall *et al.*, 1963) and nmr
(Melera, 1963) data produced a revised molecular formula of $C_{33}H_{59}O_8$
with an additional quartenary methyl group. Definitive structural work by
several groups (Halsall *et al.*, 1966; Oxley, 1966; Chou *et al.*, 1967, 1969) led
to structure **92** for cephalosporin P_1 closely related to helvolic (**93**) and

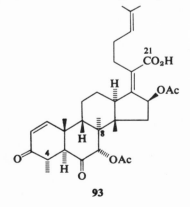

93

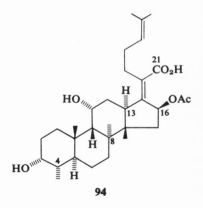

94

fusidic (**94**) acids, also products of microorganisms. All three compounds have in common a dienic side chain that includes a carboxyl group at C-21; oxygenation at C-16; lacking methyl groups at C-4 and C-13; and a rare methyl group at C-8.

Table 1.3 lists the characterized triterpenoids.

TABLE 1.3
MARINE TRITERPENOIDS

Text no.	Name	mp (in degrees)	$[\alpha]_D°$	Reference
63a	22,25-Oxidoholothurino-genin	315–316	−21.2	Chanley *et al.* (1966)
63b	17-Deoxy-22,25-oxido-holothurinogenin	286	−22	Chanley *et al.* (1966)
69	Griseogenin (22-hydroxy-holothurinogenin)	285–287	−22	Tursch *et al.* (1967)
73a	Seychellogenin (17-deoxy-holothurinogenin)	234–238	−7	Roller *et al.* (1969)
73b	Koellikerigenin (25-hydroxy-17-deoxyholo-thurinogenin)	213–214	−8	Roller *et al.* (1969)
73c	Ternaygenin (25-meth-oxy-17-deoxyholo-thurinogenin)	239–242	+2	Roller *et al.* (1969)
74	Praslinogenin (25-meth-oxyholothurinogenin)	290–291.5	—	Tursch *et al.* (1970)
75	Holothurinogenin	273	—	Habermehl and Volkwein (1970)
83	Stichopogenin A_2	238–240	48	Elyakov *et al.* (1969)
84	Stichopogenin A_4	238–240	—	Elyakov *et al.* (1969)
86	Squalene	—	—	Tsujimoto (1916)
89	Friedelin	260–262	−28.4	Tsuda and Sakai (1960)
90	Taraxerol	278	—	Santos and Doty (1971)

D. Carotenoids

As one proceeds from sesquiterpenoids via di- and triterpenoids to the tetraterpenoids, and as one looks at the structural variations among these polyisoprenoids of increasing size, one is struck by the fact that these compounds as a group become structurally more uniform as they become larger. As a result of this progression toward lesser diversity we arrive at this class of tetraterpenoids, the carotenoids, and find a coherent and well-defined group of compounds, most of which consist of a central branched nonaene chain of 22 carbon atoms with various nine-carbon moieties at each end of the central chain. Or, from an isoprenoid point of view, one can describe the carotenoids as compounds that are constructed of two diterpenoid chains connected tail to tail, with both chains being made up of four isoprene units joined head to tail. Since carotenoids possess these common structural features and since these compounds, unlike many other terpenoids, have biochemical significance that is understood, carotenoid chemistry represents a well-defined area of terpenoid chemistry that is intensively studied by relatively few research groups and the literature of which is characterized by timely reviews. A recent general review of the field is that by Davis (1968). Isler's (1971) comprehensive monograph is the standard reference for all aspects of carotenoid research.

The carotenoids are among the most widely distributed naturally occurring organic pigments, generally yellow or red in color, depending on the length and stereochemistry of the conjugated polyene chain. In spite of the relative structural uniformity of the carotenoids mentioned earlier, their isolation and structural elucidation have been formidable tasks: They occur in low concentrations, usually 0.02–0.1% of dry weight of extracted material, and normally as mixtures; they are sensitive to light, heat, and air; and they must be isolated from fresh material.

In a recent progress report on carotenoid chemistry Liaaen-Jensen (1971) plotted the number of known naturally occurring carotenoids versus the year beginning at 1900; the resulting curve is virtually smooth and exponential. The steepest slope of the curve was reached less than ten years ago, approximately coinciding with the wide use of mass spectrometry. Not surprisingly, the rapid discovery of novel structural features has taken place only during the past few years since the full development of many physical methods of structure elucidation. At the time of Liaaen-Jensen's review (1971) about three hundred natural carotenoids were recognized, about two thirds of them with correct structures assigned.

Carotenoids are generally subdivided (Davis, 1968) into the hydrocarbons or carotenes; the xanthophylls, carotenoids that possess oxygen functions other than carboxyl groups; the natural xanthophyll esters, often of palmitic acid; and the carotenoid acids. An additional group of compounds that possess fewer than forty carbon atoms and are considered degraded carotenoids are referred to as the apocarotenoids. And finally, there have been isolated a number of carotenoids with one or two extra isoprene units, but so far all of these have been found in microorganisms. Our discussion will follow this generally accepted outline; there is, however, no need from a structural viewpoint to treat the xanthophyll esters apart from the xanthophylls.

1. CAROTENES

To the nonspecialist carotene is the pigment that is responsible for the color of carrots and which is the biochemical precursor of vitamin A. Appropriately enough, when the crystalline coloring matter of carrots was first isolated in 1831 it was named carotene. Much later, with refined isolation techniques, it was recognized that carrots contained two isomeric pigments. The minor constituent, an optically active substance, was eluted ahead of the optically inactive major constituent during chromatography. Because of this elution sequence the minor component was named α-carotene and the major component β-carotene. β-Carotene (**95**) has emerged as the most abundant and as one of the most widespread of the carotenes. The standard numbering system, which follows the isoprenoid construction of the molecule, is shown in formula **95**.

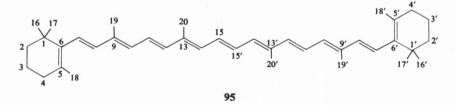

95

In contrast to the wide distribution of β-carotene (**95**), ε-carotene (**96**) is a rarely encountered hydrocarbon; it differs from the common β-isomer by having α-ionone (**97**) instead of β-ionone (**98**) end groups. ε-Carotene was first isolated from the colonial marine diatom *Navicula torquatum* by Strain and Manning (1943), and was subsequently detected in a green alga, *Bryopsis corticulans* (Strain, 1951). Chapman and Haxo (1963) reisolated ε-carotene from both marine sources as well as from the cultured unicellular flagellate *Cryptomonas ovata* var. *palustris* and showed it to be identical with a carotene

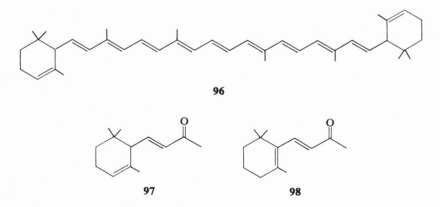

96

97   **98**

designated ε_1, that had been synthesized by Karrer and Eugster (1950). Improved syntheses have been reported by the Weedon group (Manchand *et al.*, 1965).

Carotenes with benzenoid end groups are rare and have so far been detected only in marine sources and in microorganisms. The first such compounds, a group of three closely related isomers, were first isolated from the sponge *Reniera japonica* (syn. *Halychondria panicea*) and their structures elucidated by Yamaguchi (1957a,b; 1958a,b), who also synthesized them (Yamaguchi, 1959, 1960). Yamaguchi was reluctant to accept the unprecedented aromatic end groups, but eventually yielded to convincing degradative evidence, including permanganate oxidation to trialkylbenzaldehydes. The three hydrocarbons, isolated from the sponge in relative ratios of 20:10:1 were designated renieratin (**99**), isorenieratin (**100**), and renierapurpurin (**101**). The structures were proven by synthesis, first of isorenieratin (**100**) (Yamaguchi, 1959) then of renieratin (**99**), and of renierapurpurin (**101**). More convenient and higher yield syntheses were reported by the Weedon group (Cooper *et al.*, 1963).

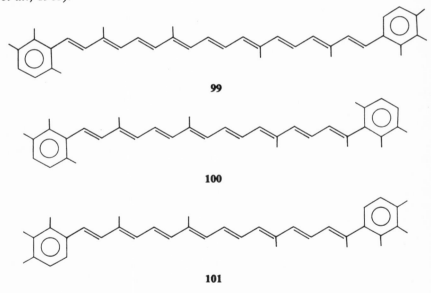

99

100

101

2. XANTHOPHYLLS

The xanthophylls are carotenoids possessing oxygen functions other than carboxyl groups. At present the xanthophylls comprise the largest number of carotenoids to have been isolated from marine organisms. Our discussion will be roughly along lines of increasing structural complexity.

One of the simplest of the oxygenated carotenoids isolated from marine sources was also one of the earliest. Lederer (1935), one of the pioneers of carotenoid chemistry, reported the isolation of echinenone (**102**) from the

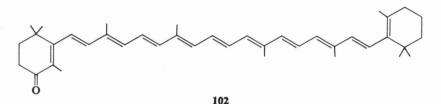

102

gonads of the sea urchin *Paracentrotus lividus* (syn. *Strongylocentrotus lividus*)* and recognized it to be a monoketone. Goodwin and Taha (1950) first suspected and later (1951) showed the pigment, which they had isolated from other marine invertebrates, to be identical with another carotenoid, myxoxanthin. Myxoxanthin was first reported by Heilbron and co-workers (1935) from the blue-green alga *Rivularia nitida* and was later characterized (Heilbron and Lythgoe, 1936) after its isolation from a freshwater blue-green alga, *Oscillatoria rubrescens*. The molecular structure of echinenone (**102**) was conclusively demonstrated by Ganguly *et al.* (1956) following its reisolation from the echinoids *Strongylocentrotus purpurata* and *S. franciscanus*. A number of syntheses of echinenone (**102**) have been reported, particularly by the Weedon group (Warren and Weedon, 1958; Akhtar and Weedon, 1959).

The major carotenoid in many marine crustaceans and asteroids is astaxanthin (**103**), which may occur as such, as a diester (e.g., a palmitate), or as a protein complex. Perhaps best known is the astaxanthin–protein complex, which is responsible for the color of the lobster shell. On cooking, the protein is denatured and the lobster shell assumes the red color of astaxanthin. Astaxanthin (**103**) was first isolated by Kuhn and Lederer (1933) from the lobster *Astacus gammarus* and was believed to be an ester because it was transformed into astacene (or astacin) (**104**) by treatment with base in the

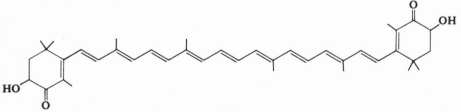

103

---

* Lederer (1935) first named the animal *Echinus esculentus*; he corrected his error in his review of invertebrate carotenoids (Lederer, 1938).

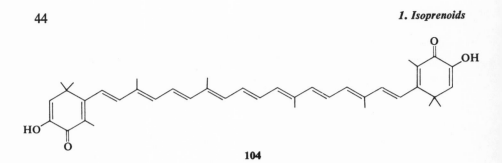

104

presence of air. Kuhn and Sørensen (1938) explained this transformation as an alkaline oxidation of an α-ketol to an α-diketone that, according to more recent data (Davis and Weedon, 1960), is enolized in solution. Both astaxanthin (**103**) and astacene (**104**) have been synthesized by Weedon and collaborators (Davis and Weedon, 1960; Leftwick and Weedon, 1967). Astaxanthin (**103**) has also been recognized as a pigment in the gonads of the sea cucumbers *Holothuria leucospilota* (Matsuno *et al.*, 1969) and *Stichopus japonicus* (Matsuno and Ito, 1971).

A closely related carotenoid, α-doradecin (**105**), was recently isolated from the goldfish *Crassius auratus*, where it occurs as the ester (α-doradexanthin ester). Its structure was secured by spectral comparison and by sodium borohydride reduction to, and ultraviolet spectral comparison with, lutein

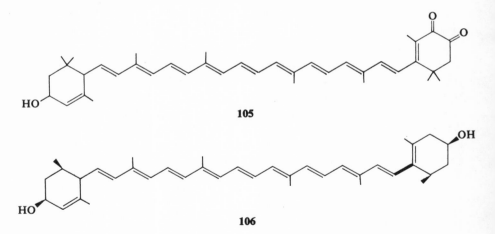

105

106

(**106**) by Katayama and co-workers (1970a,b). The authors suggest that α-doradecin may be a metabolic intermediate in the oxidation pathway of the plant carotenoids β-carotene (**95**) and lutein (**106**) to the highly oxidized animal carotenoid astacene (**104**).

Two structurally related xanthophylls, the common feature of which is an oxygenated C-9 methyl group, have been isolated from several species of green algae. Loroxanthin (**107**) was isolated from cultures of *Scenedesmus obliquus* and *Chlorella vulgaris* and shown to be a constituent in the marine species *Cladophora trichotoma, C. ovoidea,* and *Ulva rigida* (Aitzetmüller *et al.,* 1969). A probable structure (**107**) was assigned to loroxanthin by Aitzetmuller *et al.* (1969) largely on the basis of spectral data. This structure was confirmed by Walton *et al.* (1970) by reisolation and degradation. These same authors (Walton *et al.,* 1970) also elucidated structure **108** for the related siphonaxanthin on the basis of spectral data and chemical transformations. Walton *et al.* (1970) isolated siphonaxanthin from the green alga *Codium fragile.* A few years earlier, Kleinig and Egger (1967) had isolated and characterized the carotenoid following its isolation from *Caulerpa prolifera.* The German workers (Kleinig *et al.,* 1969) had determined the correct structure of siphonaxanthin (**108**) independently by chemical degradation, prior to publication of the paper by Goodwin and collaborators (Walton *et al.,* 1970).

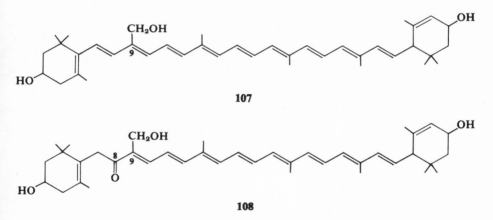

107

108

An interesting group of xanthophylls possesses acetylenic or allenic functions within the central chain and adjacent to one or both end groups. The more frequently occurring acetylenes will be discussed first.

As Weedon (1970) has pointed out in his recent comprehensive review of allenic and acetylenic carotenoids, early reports of naturally occurring acetylenic substances date back to the last century. However, until quite recently, all of the known naturally occurring acetylenes have been derivatives of unbranched carbon compounds. The first acetylenic terpenoids were reported in 1966 (Massy-Westrop *et al.,* 1966; Nozoe *et al.,* 1966) and the first acetylenic tetraterpenoids were characterized in the following year (Mallams *et al.,* 1967; Campbell *et al.,* 1967)

Chapman (1966) had isolated from several members of the algal class Cryptophyceae (*Cryptomonas ovata* var. *palustris, Rhodomonas* Strain D3, and *Hemiselmis virescens*) three apparently new carotenoids; Mallams *et al.* (1967) isolated alloxanthin (**109**) as the principal xanthophyll. The visible absorption spectrum and the chromatographic behavior of alloxanthin (**109**) were very similar to those of the widely distributed zeaxanthin (**110**), but the two compounds differed in melting points and in their IR and nmr spectra, thus suggesting a symmetrical acetylenic analog of zeaxanthin (**110**). The close similarity in the visible spectra of the two compounds, which is initially surprising, may be rationalized by two compensating parameters— the hypsochromic tendency of the triple bonds in alloxanthin is counteracted by the bathochromic effect caused by the ready planarity of the conjugated system that is engendered by the acetylenic linkages (Weedon, 1970). Alloxanthin furthermore was shown to be identical with carotenoids that had been described earlier—pectenoxanthin from the giant scallop *Pecten maximus* and cynthiaxanthin from the tunicate *Halocynthia papillosa* (Campbell *et al.*, 1967). It has also been isolated from the mussel *Mytilus edulis*.

The two minor xanthophylls from the same algal source (Chapman, 1966) that yielded alloxanthin (**109**) were shown to be monadoxanthin of structure **111** and crocoxanthin, **112** (Mallams *et al.*, 1967). Campbell and co-workers (1967) isolated another new acetylenic carotenoid from the gonads of the scallop *Pecten maximus*. They named the compound pectenolone and showed its structure to be **113**. This compound was also a minor pigment in the tunicate *Halocynthia papillosa* (Campbell *et al.*, 1967).

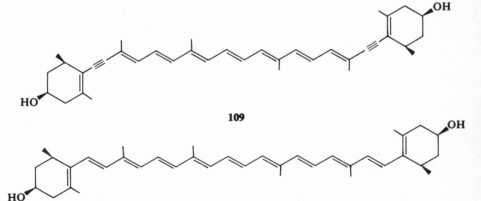

109

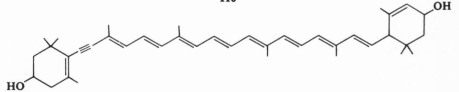

110

111

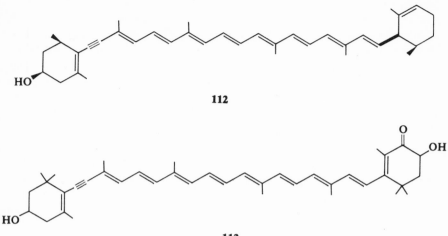

112

113

Another pair of xanthophylls that proved to have acetylenic linkages were first isolated by Strain and co-workers (1944) from a number of diatoms including *Navicula torquatum* and *Isthmia nervosa*. Strain *et al.* (1944) designated the new pigments diatoxanthin (**114**) and diadinoxanthin (**115**) and noted that the visible absorption spectra of the two carotenoids strongly

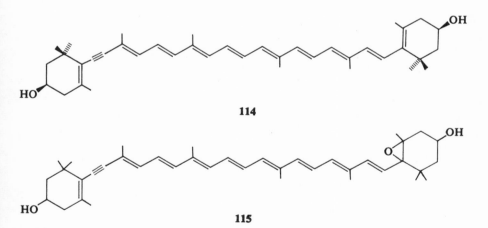

114

115

resembled those of zeaxanthin (**110**) and of lutein (**106**), respectively. Chapman (1966) isolated diatoxanthin (**114**) from *Isochrysis galbana* and Mallams *et al.* (1967) assigned structure **114** to the pigment on the basis of spectral comparisons. The companion xanthophyll of diatoms, diadinoxanthin, which has also been isolated from dinoflagellates and from *Euglena gracilis*, is the epoxide of diatoxanthin (**114**) and has structure **115** (Aitzetmüller *et al.*, 1968).

A mixture of acetylenic carotenoids that are related to, and were previously believed to be identical with, astaxanthin (**103**) had been isolated from the sea star *Asterias rubens* and was known in the literature as asterinic acid. Recent reinvestigation by the Norwegian group (Sørensen *et al.*, 1968; Liaaen-Jensen, 1969) showed asterinic acid to be mono- and diacetylenic analogs of astaxanthin (**103**) of structures **116** and **117**.

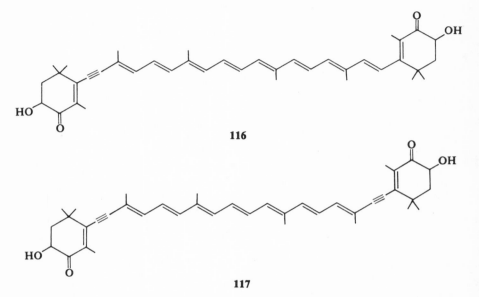

116

117

Naturally occurring allenic substances are far fewer in number than are acetylenes. The first authentic example is that of mycomycin (**118**), a linear C_{13} compound that is an antibiotic isolated from a fungus (Celmer and Solomons, 1952a,b). Almost all naturally occurring allenes have been isolated from microorganisms and they have unbranched carbon skeletons (Jones, 1966). In view of this backdrop it is rather amazing that the principal carotenoid of brown algae (Phaeophyceae) and perhaps the most abundant natural

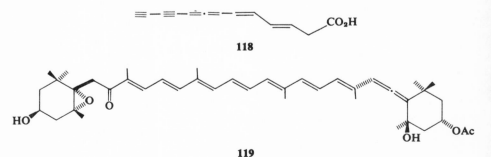

118

119

carotenoid (Allen *et al.*, 1960; Bonnett *et al.*, 1964, 1969), fucoxanthin (**119**) is a representative of this exclusive class of natural products. Not surprisingly, however, the structure of fucoxanthin, an investigation dating back to 1914, has been a difficult problem that was recently solved by the Weedon group (Bonnett *et al.*, 1964), who have also provided the historical background and full experimental details in what doubtless is a classic paper in the carotenoid field (Bonnett *et al.*, 1969). Among the outstanding features of this investigation was the recognition of the 1927 cm^{-1} ir band as that of an allene. The acetate, novel in a carotenoid, was unambiguously identified by the properties of perhydrofucoxanthin—appropriate ir absorption, O-acetyl analysis, and hydrolysis yielding acetic acid. High-resolution mass spectrometry provided an unambiguous molecular formula of $C_{42}H_{58}O_6$ for fucoxanthin, thereby confirming its nature as an acetate. Many of the structural details were pieced together by the use of zinc rather than the conventional potassium permanganate as a principal oxidizing agent. The zinc salt reacted faster than the potassium analog and caused fewer side reactions, because of its weakly basic nature. It allowed the isolation and structural elucidation of, *inter alia*, authentic allenic degradation products. Needless to say, the monumental task of solving the fucoxanthin structure could not have been accomplished without the extensive and imaginative use of nuclear magnetic resonance and high resolution mass spectrometric techniques.

Stereochemical assignments, including absolute stereochemistry of this fascinating molecule, have been achieved by the chemical conversion of fucoxanthin (**119**) into zeaxanthin (**110**) (Bonnett *et al.*, 1969), and by X-ray diffraction analysis of a *p*-bromobenzoate of an allenic degradation product of fucoxanthin (**120**) (De Ville *et al.*, 1969).

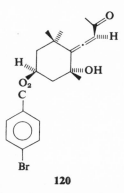

120

Lederer (1935, 1938) in his early work on the pigments of marine invertebrates had noted the presence of uncommon carotenoids in sea urchins. Recent reinvestigation (Galasko *et al.*, 1969) of the carotenoids of the sea urchin *Paracentrotus lividus* has led to the isolation of a trace of fucoxanthin

(119) in addition to fucoxanthinol (121) and the C_{31} carotenoid paracentrone (122).

Fucoxanthinol (121), the parent alcohol of the acetate fucoxanthin (119), had not previously been observed as a natural product and cannot be prepared *in vitro* by direct acid or base hydrolysis of fucoxanthin (119) (Bonnett

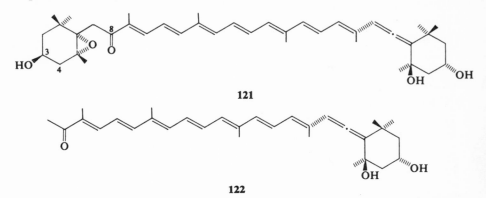

121

122

et al., 1969). It had, however, been made indirectly from fucoxanthin (119) by selective reduction of the acetate and the C-8 oxo function and subsequent reoxidation of the newly generated C-8 allylic alcohol (Bonnett *et al.*, 1966, 1969). The third pigment, paracentrone (122), proved to be one of the earliest examples of an apocarotenoid.

3. APOCAROTENOIDS

The structure of this unusual C_{31} pigment, paracentrone (122), was assigned on the basis of spectral data (Galasko *et al.*, 1969). The authors suggest that both fucoxanthinol (121) and paracentrone (122) are derived from dietary fucoxanthin (119) and are degraded by the sea urchin. A laboratory conversion of fucoxanthin (119) to paracentrone (122) has been achieved (Hora *et al.*, 1970) by Oppenauer oxidation of fucoxanthin directly to paracentrone acetate, from which paracentrone (122) was prepared by alkaline hydrolysis. The successful *in vitro* degradation of fucoxanthin (119) to paracentrone (122) lends support to the authors' (Hora *et al.*, 1970) suggestion that *in vivo* transformation may also be triggered by oxidation of the C-3 hydroxy group, thereby creating an active methylene function at C-4.

The structure of another allenic apocarotenoid, peridinin (123), has recently been fully elucidated in a collaborative effort by four research groups (Strain *et al.*, 1971). Peridinin (123) is the principal carotenoid pigment of dinoflagellates, marine and freshwater, and was first isolated in 1890 (Schütt, 1890) from the marine genera *Ceratium* and *Peridinium*. It has also been isolated from the zooxanthellae of corals, clams, and sea anemones.

For the successful structural work peridinin (**123**) was isolated from the zooxanthellae of the sea anemone *Bunodactis* (syn. *Anthopleura*) *xanthogrammica*, from a bloom of the dinoflagellate *Gonyaulax polyedra*, and from cultures of *Cachonina hiei* and *Amphidinium operculatum*. Chemical degradation and extensive spectral analysis led to the assignment of structure **123** for peridinin, which is the natural acetate of a C_{37} carotenoid. Peridinin (**123**) is found in the organisms along with normal C_{40} carotenoids and no doubt represents a carotenoid degradation product, albeit a more complex one than paracentrone (**122**), in that the carotenoid end groups are intact and degradation has taken place in the central carbon chain. Along with fucoxanthin (**119**) peridinin (**123**) appears to be a carotenoid of fundamental photosynthetic significance (Haxo, 1960).

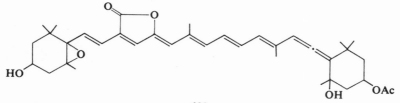

123

Another new class of apocarotenoids is represented by actinioerythrin (**124b**), the principal carotenoid pigment of the sea anemone *Actinia equina*. It was first isolated by Lederer (1933), who recognized it as a xanthophyll ester. Successful structural elucidation was achieved by chemical and spectral techniques (Hertzberg and Liaaen-Jensen, 1968; Hertzberg *et al.*, 1969). The parent alcohol, actinioerythrol (**124a**), is esterified in nature with C_{10}, C_{11}, C_{12}, and perhaps other fatty acids (Hertzberg *et al.*, 1969). The alcohol

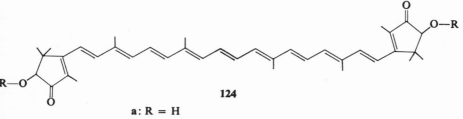

124

a: R = H
b: R = mixture of C_{10}–C_{12} acyl groups

itself (**124a**) is a remarkable bisnorcarotenoid with ring-contracted cyclopentenone end groups. The authors (Hertzberg *et al.*, 1969) have postulated that *in vivo* biogenesis may originate from astaxanthin (**103**) via a triketone followed by benzilic acid rearrangement and decarboxylation. The *in vitro* feasibility of such a transformation has been demonstrated by Holzel and co-workers (1969).

TABLE 1.4
MARINE CAROTENOIDS[a]

Text no.	Name	Composition	mp (in degrees)	Reference
96	ε-Carotene	$C_{40}H_{56}$	198–199	Manchand *et al.* (1965)
99	Renieratin	$C_{40}H_{48}$	185	Yamaguchi (1957a)
100	Isorenieratin	$C_{40}H_{48}$	199	Yamaguchi (1957a)
101	Renierapurpurin	$C_{40}H_{48}$	230	Yamaguchi (1957a)
102	Echinenone (myxoxanthin)	$C_{40}H_{54}O$	179–181	Ganguly *et al.* (1956)
103	Astaxanthin	$C_{40}H_{52}O_4$	216	Kuhn and Sørensen (1938)
105	α-Doradecin	$C_{40}H_{50}O_2$	144–156	Katayama *et al.* (1970a)
107	Loroxanthin	$C_{40}H_{56}O_3$	—	Aitzetmüller *et al.* (1969)
108	Siphonaxanthin	$C_{40}H_{56}O_4$	—	Kleinig and Egger (1967)
109	Alloxanthin (pectenoxanthin, cynthiaxanthin)	$C_{40}H_{52}O_2$	186–188	Chapman (1966)
111	Monadoxanthin	$C_{40}H_{54}O_2$	165	Chapman (1966)
112	Crocoxanthin	$C_{40}H_{54}O$	163–165	Chapman (1966)
113	Pectenolone	$C_{40}H_{52}O_3$	—	Campbell *et al.* (1967)
114	Diatoxanthin	$C_{40}H_{54}O_2$	201	Chapman (1966)
115	Diadinoxanthin	$C_{40}H_{54}O_3$	158–162	Aitzetmüller *et al.* (1968)
119	Fucoxanthin	$C_{42}H_{58}O_6$	168–169	Bonnett *et al.* (1969)
121	Fucoxanthinol	$C_{40}H_{56}O_5$	146–148	Galasko *et al.* (1969)
122	Paracentrone	$C_{31}H_{42}O_3$	147–149	Galasko *et al.* (1969)
123	Peridinin	$C_{39}H_{50}O_7$	107–109	Strain *et al.* (1971)
124a	Actinioerythrol	$C_{38}H_{48}O_4$	[b]	Hertzberg and Liaaen-Jensen (1968)

[a] A majority of carotenoids are optically inactive. Few rotation values appear to be recorded for those carotenoids that possess optical activity.

[b] The natural diester, actinioerythrin, has mp 91°.

REFERENCES

Aitzetmüller, K., Svec, W. A., Katz, J. J., and Strain, H. H. (1968). *Chem. Commun.* 32.
Aitzetmüller, K., Strain, H. H., Svec, W. A., Grandolfo, M., and Katz, J. J. (1969). *Phytochemistry* **8**, 1761.
Akhtar, M., and Weedon, B. C. L. (1959). *J. Chem. Soc.* 4058.
Allen, F. H., and Rogers, D. (1966). *Chem. Commun.* 837.
Allen, M. B., Goodwin, T. W., and Phagpolngarm, S. (1960). *J. Gen. Microbiol.* **23**, 93.
Baslow, M. H. (1969). "Marine Pharmacology," pp. 17–18. Williams & Wilkins, Baltimore, Maryland.

Bates, R. B., Büchi, G., Matsuura, T., and Shaffer, R. R. (1960). *J. Amer. Chem. Soc.* **82**, 2327.

Beaton, J. M., Spring, F. S., Stevenson, R., and Stewart, J. L. (1954). *Chem. Ind.(London)* 1454.

Beaton, J. M., Spring, F. S., Stevenson, R., and Stewart, J. L. (1955a). *Chem. Ind. (London)* 35.

Beaton, J. M., Spring, F. S., Stevenson, R., and Stewart, J. L. (1955b). *J. Chem. Soc.* 2131.

Blumer, M., Mullin, M. M., and Thomas, D. W. (1963). *Science* **140**, 974.

Bonnett, R., Spark, A. A., Tee, J. L., and Weedon, B. C. L. (1964). *Proc. Chem. Soc.* 419.

Bonnett, R., Mallams, A. K., McCormick, A., Tee, J. L., and Weedon, B. C. L. (1966). *Chem. Commun.* 515.

Bonnett, R., Mallams, A. K., Spark, A. A., Tee, J. L., Weedon, B. C. L., and McCormick, A. (1969). *J. Chem. Soc. C.* 429.

Bryant, R. (1969). *In* "Rodd's Chemistry of Carbon Compounds" (S. Coffey, ed.), 2nd ed., Vol. IIC, p. 257. Elsevier, Amsterdam.

Burton, H. S., Abraham, E. P., and Cardwell, H. M. (1956). *Biochem. J.* **62**, 171.

Cafieri, F., Fattorusso, E., Santacroce, C., and Minale, L. (1972). *Tetrahedron* **28**, 1579.

Cameron, A. F., Ferguson, G., and Robertson, J. M. (1967). *Chem. Commun.* 271.

Cameron, A. F., Ferguson, G., and Robertson, J. M. (1969). *J. Chem. Soc. B* 692.

Campbell, S. A., Mallams, A. K., Waight, E. S., Weedon, B. C. L., Barbier, M., Lederer, E., and Salaque, A. (1967). *Chem. Commun.* 941.

Carboni, S., Da Settino, A., Malaguzzi, V., Marsili, A., and Pacini, P. L. (1965). *Tetrahedron Lett.* 3017.

Celmer, W. D., and Solomons, I. A. (1952a). *J. Amer. Chem. Soc.* **74**, 1870.

Celmer, W. D., and Solomons, I. A. (1952b). *J. Amer. Chem. Soc.* **74**, 2245.

Chanley, J. D., and Rossi, C. (1969a). *Tetrahedron* **25**, 1897.

Chanley, J. D., and Rossi, C. (1969b). *Tetrahedron* **25**, 1911.

Chanley, J. D., Kohn, S. K., Nigrelli, R. F., and Sobotka, H. (1955). *Zoologica* **40**, 99.

Chanley, J. D., Ledeen, R., Wax, J., Nigrelli, R. F., and Sobotka, H. (1959). *J. Amer. Chem. Soc.* **81**, 5180.

Chanley, J. D., Perlstein, J., Nigrelli, R. F., and Sobotka, H. (1960). *Ann. N.Y. Acad. Sci.* **90**, 902.

Chanley, J. D., Mezzetti, T., and Sobotka, H. (1966). *Tetrahedron* **22**, 1857.

Chapman, D. J. (1966). *Phytochemistry* **5**, 1331.

Chapman, D. J., and Haxo, F. T. (1963). *Plant Cell Physiol.* **4**, 57.

Christensen, P. K., and Sørensen, N. A. (1951). *Acta Chem. Scand.* **5**, 751.

Chou, T. S., Eisenbraun, E. J., and Rapala, R. T. (1967). *Tetrahedron Lett.* 409.

Chou, T. S., Eisenbraun, E. J., and Rapala, R. T. (1969). *Tetrahedron* **25**, 3341.

Ciereszko, L. S. (1962). *Trans. N.Y. Acad. Sci.* **24**, Ser. 2, 502 (1962).

Ciereszko, L. S., Sifford, D. H., and Weinheimer, A. J. (1960). *Ann. N.Y. Acad. Sci.* **90**, 917.

Cimino, G., DeStefano, S., Minale, L., and Fattorusso, E. (1972). *Tetrahedron* **28**, 333.

Cooper, R. D. J., Davis, J. B., and Weedon, B. C. L. (1963). *J. Chem. Soc.* 5637.

Corey, E. J., and Ursprung, J. J. (1955a). *J. Amer. Chem. Soc.* **77**, 3667.

Corey, E. J., and Ursprung, J. J. (1955b). *J. Amer. Chem. Soc.* **77**, 3668.

Crawford, K., Heatley, N. G., Boyd, P. F., Hale, C. W., Kelly, B. K., Miller, G. A., and Smith, N. (1952). *J. Gen. Microbiol.* **6**, 47.

Davis, J. B. (1968). *In* "Rodd's Chemistry of Carbon Compounds" (S. Coffey, ed.), 2nd ed., Vol. IIB, pp. 231–346. Elsevier, Amsterdam.

Davis, J. B., and Weedon, B. C. L. (1960). *Proc. Chem. Soc.* 182.

de Mayo, P., Williams, R. E., Büchi, G., and Feairheller, S. H. (1965). *Tetrahedron* **21**, 619.

De Ville, T. E., Hursthouse, M. B., Russell, S. W., and Weedon, B. C. L. (1969). *J. Chem. Soc. D.* 1311.

Dixit, S. M., Rao, A. S., and Paknikar, S. K. (1967). *Chem. Ind. (London)* 1256.

Dutler, H., Jeger, O., and Ruzicka, L. (1955). *Helv. Chim. Acta* **38**, 1268.

Elyakov, G. B., Kuznetsova, T. A., Dzizenko, A. K., and Elkin, Yu. N. (1969). *Tetrahedron Lett.* 1151.

Enzell, C. and Erdtman, H. (1958). *Tetrahedron* **4**, 361.

Erdtman, H., Kimland, B., Norin, T., and Daniels, P. J. L. (1968). *Acta Chem. Scand.* **22**, 938.

Fattorusso, E., Minale, L., Sodano, G., and Trivellone, E. (1971). *Tetrahedron* **27**, 3909.

Fenical, W., Sims, J. J., Wing, R. M., and Radlick, P. (1971). *Abstracts, Pacific Conference on Chemistry and Spectroscopy*, Anaheim, Cal., p. 35.

Friedel, C. (1892). *Bull. Soc. Chim. Fr.* (3) **7**, 164.

Friess, S. L., Durant, R. C., and Chanley, J. D. (1968). *Toxicon* **6**, 81.

Galasko, G., Hora, J., Toube, T. P., Weedon, B. C. L., André, D., Barbier, M., Lederer, E., and Villanueva, V. R. (1969). *J. Chem. Soc. C* 1264.

Gallagher, M. J., Hildebrand, R. P., and Sutherland, M. D. (1964). *Tetrahedron Lett.* 3715.

Ganguly, J., Krinsky, N. I., and Pinckard, J. H. (1956). *Arch. Biochem. Biophys.* **60**, 345.

Goodwin, T. W., and Taha, M. M. (1950). *Biochem. J.* **47**, 244.

Goodwin, T. W., and Taha, M. M. (1951). *Biochem. J.* **48**, 513.

Gupta, K. C., and Scheuer, P. J. (1968). *Tetrahedron* **24**, 5831.

Habermehl, G., and Volkwein, G. (1968). *Naturwissenschaften* **55**, 83.

Habermehl, G., and Volkwein, G. (1970). *Justus Liebigs Ann. Chem.* **731**, 53.

Habermehl, G., and Volkwein, G. (1971). *Toxicon* **9**, 319.

Halsall, T. G., Jones, E. R. H., and Lowe, G. (1963). *Proc. Chem. Soc.* 16.

Halsall, T. G., Jones, E. R. H., Lowe, G., and Newall, C. E. (1966). *Chem. Commun.* 685.

Halstead, B. W. (1965). "Poisonous and Venomous Marine Animals of the World," Vol. 1, p. 570. U.S. Government Printing office, Washington, D.C.

Hashimoto, Y., and Yasumoto, T. (1960). *Bull. Jap. Soc. Sci. Fish.* **26**, 1132.

Haxo, F. T. (1960). *In* "Comparative Biochemistry of Photoreactive Systems" (M. B. Allen, ed.), p. 339. Academic Press, New York.

Heilbron, I. M., and Lythgoe, B. (1936). *J. Chem. Soc.* 1376.

Heilbron, I. M., Lythgoe, B., and Phipers, R. F. (1935). *Nature* **136**, 989.

Hertzberg, S., and Liaaen-Jensen, S. (1968). *Acta Chem. Scand.* **22**, 1714.

Hertzberg, S., Liaaen-Jensen, S., Enzell, C. R., and Francis, G. W. (1969). *Acta Chem. Scand.* **23**, 3290.

Holzel, R., Leftwick, A. P., and Weedon, B. C. L. (1969). *J. Chem. Soc. D*, 128.

Hora, J., Toube, T. P., and Weedon, B. C. L. (1970). *J. Chem. Soc. C*, 241.

Horeau, A. (1961). *Tetrahedron Lett.* 506.

Horeau, A. (1962). *Tetrahedron Lett.* 965.

Houssain, M. B., and van der Helm, D. (1968). *J. Amer. Chem. Soc.* **90**, 6607.

Houssain, M. B., and van der Helm, D. (1969). *Rec. Trav. Chim. Pays-Bas* **88**, 1413.

Houssain, M. B., Nicholas, A. F., and van der Helm, D. (1968). *Chem. Commun.* 385.

Hunter, G. L. K., and Brogden, W. B., Jr. (1964). *J. Org. Chem.* **29**, 2100.

Irie, T., Yamamoto, K., and Masamune, T. (1964). *Bull. Chem. Soc. Japan* **37**, 1053.

Irie, T., Suzuki, M., and Masamune, T. (1965a). *Tetrahedron Lett.* 1091.

Irie, T., Yasunari, Y., Suzuki, T., Imai, N., Kurosawa, E., and Masamune, T. (1965b). *Tetrahedron Lett.* 3619.

Irie, T., Suzuki, M., Kurosawa, E., and Masamune, T. (1966). *Tetrahedron Lett.* 1837.

Irie, T., Suzuki, T., Itô, S., and Kurosawa, E. (1967). *Tetrahedron Lett.* 3187.

Irie, T., Fukuzawa, A., Izawa, M., and Kurosawa, E. (1969). *Tetrahedron Lett.* 1343.

Irie, T., Suzuki, M., and Hayakawa, Y., (1969). *Bull. Chem. Soc. Jap.* **42**, 843.

Irie, T., Suzuki, M., Kurosawa, E., and Masamune, T. (1970). *Tetrahedron* **26**, 3271.

Isler, O., ed. (1971). "Carotenoids." Birkhäuser Verlag, Basel, Switzerland.

Itô, S., Endo, K., Yoshida, T., Yatagai, M., and Kodama, M. (1967). *Chem. Commun.* 186.

Jones, E. R. H. (1966). *Chem. Brit.* **2**, 6.

Joshi, B. N., Seshadri, R., Chakravarti, K. K., and Bhattacharyya, S. C. (1964). *Tetrahedron* **20**, 2911.

Karrer, P., and Helfenstein, A. (1931). *Helv. Chim. Acta* **14**, 78.

Karrer, P., and Eugster, C. H. (1950). *Helv. Chim. Acta* **33**, 1433.

Karrer, P., Helfenstein, A., Wehrli, H., and Wettstein, A. (1930). *Helv. Chim. Acta* **13**, 1084.

Katayama, T. (1962). *In* "Physiology and Biochemistry of Algae" (R. A. Lewin, ed.) pp. 467–473. Academic Press, New York.

Katayama, T., Yokoyama, H., and Chichester, C. O. (1970a). *Int. J. Biochem.* **1**, 438.

Katayama, T., Yokoyama, H., and Chichester, C. O. (1970b). *Jap. Soc. Sci. Fish.* **36**, 702.

Kimland, B., and Norin, T. (1968). *Acta Chem. Scand.* **22**, 943.

Kleinig, H., and Egger, K. (1967). *Phytochemistry* **6**, 1681.

Kleinig, H., Nitsche, H., and Egger, K. (1969). *Tetrahedron Lett.* 5139.

Kuhn, R., and Lederer, E. (1933). *Chem. Ber.* **66**, 488.

Kuhn, R., and Sørensen, N. A. (1938). *Chem. Ber.* **71**, 1879.

Kurosawa, E., Izawa, M., Yamamoto, K., Masamune, T., and Irie, T. (1966). *Bull. Chem. Soc. Japan* **39**, 2509.

Lederer, E. (1933). *C. R. Soc. Biol.* **113**, 1391.

Lederer, E. (1935). *C. R. Acad. Sci. C* **201**, 300.

Lederer, E. (1938). *Bull. Soc. Chim. Biol.* **20**, 567.

Leftwick, A. P., and Weedon, B. C. L. (1967). *Chem. Commun.* 49.

Liaaen-Jensen, S. (1969). *Pure Appl. Chem.* **20**, 421.

Liaaen-Jensen, S. (1971). *In* "Aspects of Terpenoid Chemistry and Biochemistry" (T. W. Goodwin, ed.), pp. 223–254. Academic Press, New York.

Mallams, A. K., Waight, E. S., Weedon, B. C. L., Chapman, D. J., Haxo, F. T., Goodwin, T. W., and Thomas, D. M. (1967). *Chem. Commun.* 301.

Manchand, P. S., Rüegg, R., Schwieter, U., Siddons, P. T., and Weedon, B. C. L. (1965). *J. Chem. Soc.* 2019.

Manjarrez, A., and Guzmán, A. (1966). *J. Org. Chem.* **31**, 348.

Massy-Westropp, R. A., Reynolds, G. D., and Spotswood, T. M. (1966). *Tetrahedron Lett.* 1939.

Matsuda, H., Tomiie, Y., Yamamura, S., and Hirata, Y. (1967). *Chem. Commun.* 898.

Matsuno, T., and Yamanouchi, T. (1961). *Nature (London)* **191**, 75.

Matsuno, T., and Ito, T. (1971). *Experientia* **27**, 509.

Matsuno, T., Ishida, T., Ito, T., and Sakushima, A. (1969). *Experientia* **25**, 1253.

Melera, A. (1963). *Experientia* **19**, 565.

Nigrelli, R. F. (1952). *Zoologica* **37**, 89.

Nigrelli, R. F., and Jakowska, S. (1960). *Ann. N. Y. Acad. Sci.* **90**, 884.

Nigrelli, R., and Zahl, P. A. (1952). *Proc. Soc. Exp. Biol. Med.* **81**, 379.

Nigrelli, R. F., Chanley, J. D., Kohn, S. K., and Sobotka, H. (1955). *Zoologica* **40**, 47.

Nomura, T., Tsuchiya, Y., André, D., and Barbier, M. (1969a). *Bull. Jap. Soc. Sci. Fish.* **35**, 293.

Nomura, T., Tsuchiya, Y., André, D., and Barbier, M. (1969b). *Bull. Jap. Soc. Sci. Fish.* **35**, 299.

Nozoe, T., Cheng, Y. S., and Toda, T. (1966). *Tetrahedron Lett.* 3663.

Obata, Y. and Fukushi, S. (1953). *J. Agr. Chem. Soc. Japan* **27**, 331. [*Chem. Abstr.* **48**, 14127 (1954).]

Ohta, Y., and Hirose, Y. (1969). *Tetrahedron Lett.* 1601.

Ohta, Y., Sakai, T., and Hirose, Y. (1966). *Tetrahedron Lett.* 6365.

Ohta, Y., Ohara, K., and Hirose, Y. (1968). *Tetrahedron Lett.* 4181.

Ourisson, G., Crabbé, P., and Rodig, O. R. (1964). "Tetracyclic Triterpenes," p. 59. Hermann, Paris.

Oxley, P. (1966). *Chem. Commun.* 729.

Parker, W., Ramage, R., and Raphael, R. A. (1962). *J. Chem. Soc.* 1558.

Patil, L. J., and Rao, A. S. (1967). *Tetrahedron Lett.* 2273.

Piers, E., Britton, R. W., and de Waal, W. (1969). *Tetrahedron Lett.* 1251.

Pigulevskii, G. V., and Borokov, A. V. (1962). *J. Gen. Chem. USSR* **32**, 3054.

Pollock, J. R. A., and Stevens, R. Eds., (1965). "Dictionary of Organic Compounds," 4th ed., p. 1929. Oxford Univ. Press, London and New York.

Roller, P., Djerassi, C., Cloetens, R., and Tursch, B. (1969). *J. Amer. Chem. Soc.* **91**, 4918.

Roller, P., Tursch, B., and Djerassi, C. (1970). *J. Org. Chem.* **35**, 2585.

Ruzicka, L., and Stoll, M. (1922). *Helv. Chim. Acta* **5**, 929.

Ruzicka, L., Wind, A. H., and Koolhaas, D. R. (1931). *Helv. Chim. Acta* **14**, 1132.

Santos, G. A., and Doty, M. S. (1971). *Lloydia* **34**, 88.

Schütt, F. (1890). *Ber. Deut. Bot. Ges.* **8**, 9.

Shimada, S. (1969). *Science* **163**, 1462.

Sims, J. J., Fenical, W., Wing, R. M., and Radlick, P. (1971). *J. Amer. Chem. Soc.* **93**, 3774.

Sims, J. J., Fenical, W., Wing, R. M., and Radlick, P. (1972). *Tetrahedron Lett.* 195.

Singh, H., Moore, R. E., and Scheuer, P. J. (1967). *Experientia* **23**, 624.

Šorm, F., Holub, M., Sýkora, V., Mleziva, J., Streibl, M., Plíva, J., Schneider, B., and Herout, V. (1953). *Collect. Czech. Chem. Commun.* **18**, 512.

Sørensen, N. A., and Mehlum J. (1948). *Acta Chem. Scand.* **2**, 140.

Sørensen, J. S., and Sørenson, N. A. (1949). *Acta Chem. Scand.* **3**, 939.

Sørensen, N. A., Liaaen-Jensen, S., Børdalen, B., Hang, A., Enzell, C., and Francis, G. (1968). *Acta Chem. Scand.* **22**, 344.

Stevens, R., Ed. (1969). "Dictionary of Organic Compounds," 4th ed., Fifth and Cumulative Supplement, p. 106. Oxford Univ. Press, London and New York.

Strain, H. H. (1951). *In* "Manual of Phycology" (G. M. Smith, ed.), pp. 243–262. Chronica Botanica, Waltham, Massachusetts.

Strain, H. H., and Manning, W. M. (1943). *J. Amer. Chem. Soc.* **65**, 2258.

Strain, H. H., Manning, W. M., and Hardin, G. (1944). *Biol. Bull.* **86** 169.

Strain, H. H., Svec, W. A., Aitzetmüller, K., Grandolfo, M. C., Katz, J. J., Kjøsen, H., Norgård, S., Liaaen-Jensen, S., Haxo, F. T., Wegfahrt, P., and Rapoport, H. (1971). *J. Amer. Chem. Soc.* **93**, 1823.

Streith, J., Pesnelle, P., and Ourisson, G. (1963). *Bull. Soc. Chim. Fr.* 518.

Suzuki, M., Hayakawa, Y., and Irie, T. (1969). *Bull. Chem. Soc. Japan* **42**, 3342.

Suzuki, M., Kurosawa, E., and Irie, T. (1970). *Tetrahedron Lett.* 4995.

Sýkora, V., Herout, V., and Šorm, F. (1956). *Collect. Czech. Chem. Commun.* **21**, 267.

Takaoka, M., and Ando, Y. (1951). *J. Chem. Soc. Japan, Pure Chem. Sec.* **72**, 999. [*Chem. Abstr.* **46**, 9261 (1952).]

Tanaka, A., Uda, H., and Yoshikoshi, A. (1967). *Chem. Commun.* 188.

Tanaka, A., Uda, H., and Yoshikoshi, A. (1969). *Chem. Commun.* 308.

Tanaka, T., and Toyama, Y. (1959). *J. Chem. Soc. Japan, Pure Chem. Sec.* **80**, 1329.

Trivedi, G. K., Wagh, A. D., Paknikar, S. K., Chakravarti, K. K., and Bhatacharyya, S. C. (1966). *Tetrahedron* **22**, 1641.

Tsuda, K., and Sakai, K. (1960). *Chem. Pharm. Bull.* (Tokyo) **8**, 554.

Tsujimoto, M. (1916). *J. Chem. Ind. Japan* **19**, 277. [*Chem. Abstr.* **10**, 1602 (1916).]

Tsujimoto, M. (1917). *J. Ind. Eng. Chem.* **9**, 1098. [*Chem. Abstr.* **12**, 223 (1918).]

Tsujimoto, M. (1935). *Bull. Chem. Soc. Japan* **10**, 149.

Turner, A. B., Smith, D. S. H., and Mackie, A. M. (1971). *Nature* **233**, 209.

Tursch, B., de Souza Guimarães, I. S., Gilbert, B., Aplin, R. T., Duffield, A. M., and Djerassi, C. (1967). *Tetrahedron* **23**, 761.

Tursch, B., Cloetens, R., and Djerassi, C. (1970). *Tetrahedron Lett.* 467.

Vrkoč, V., Křepinský, J., Herout, V., and Šorm, F. (1964). *Collect. Czech. Chem. Commun.* **29**, 795.

Wallach, O. (1887). *Justus Liebigs Ann. Chem.* **238**, 78.

Walton, T. J., Britton, G., Goodwin, T. W., Diner, B., and Moshier, S. (1970). *Phytochemistry* **9**, 2545.

Warren, C. K., and Weedon, B. C. L. (1958). *J. Chem. Soc.* 3986.

Weedon, B. C. L. (1970). *Rev. Pure Appl. Chem.* **20**, 51.

Weinheimer, A. J., and Washecheck, P. H. (1969). *Tetrahedron Lett.* 3315.

Weinheimer, A. J., Middlebrook, R. E., Bledsoe, J. O., Jr., Marsico, W. E., and Karns, T. K. B. (1968). *Chem. Commun.* 384.

Weinheimer, A. J., Schmitz, F. J., and Ciereszko, L. S. (1968). *In* "Drugs from the Sea" (H. D. Freudenthal, ed.), p. 135. Marine Technology Society, Washington, D.C.

Weinheimer, A. J., Washecheck, P. H., van der Helm, D., and Houssain, M. B. (1968). *Chem. Commun.* 1070.

Weinheimer, A. J., Youngblood, W. W., Washecheck, P. H., Karns, T. K. B., and Ciereszko, L. S. (1970). *Tetrahedron Lett.* 497 (1970).

Winkler, L. R., Tilton, B. E., and Hardinge, M. G. (1962). *Arch. Intern. Pharmacodyn.* **137**, 76.

Yamada, K., Yazawa, H., Toda, M., and Hirata, Y. (1968). *Chem. Commun.* 1432.

Yamaguchi, M. (1957a). *Bull. Chem. Soc. Japan* **30**, 111.

Yamaguchi, M. (1957b). *Bull. Chem. Soc. Japan* **30**, 979.

Yamaguchi, M. (1958a). *Bull. Chem. Soc. Japan* **31**, 51.

Yamaguchi, M. (1958b). *Bull. Chem. Soc. Japan* **31**, 739.

Yamaguchi, M. (1959). *Bull. Chem. Soc. Japan* **32**, 1171.

Yamaguchi, M. (1960). *Bull. Chem. Soc. Japan* **33**, 1560.

Yamamura, S., and Hirata, Y. (1963). *Tetrahedron* **19**, 1485.

Yamamura, S., and Hirata, Y. (1971). *Bull. Chem. Soc. Japan* **44**, 2560 (1971).

Yamanouchi, T. (1942). *Teikoku Gakushiin Hokôku*, 73.

Yamanouchi, T. (1955). *Publ. Seto Mar. Biol. Lab.* **4**, 183.

Yasumoto, T., and Hashimoto, Y. (1965). *Agr. Biol. Chem.* (*Tokyo*) **29**, 804.

Yasumoto, T., and Hashimoto, Y. (1967). *Agr. Biol. Chem.* (*Tokyo*) **31**, 638.

Yasumoto, T., Tanaka, K., and Hashimoto, Y. (1966). *Bull. Jap. Soc. Sci. Fish.* **32**, 673.

Yasumoto, T., Nakamura, K., and Hashimoto, Y. (1967). *Agr. Biol. Chem.* (*Tokyo*) **31**, 7.

Zabża, A., Romaňuk, M., and Herout, V. (1966). *Collect. Czech. Chem. Commun.* **31**, 3373.

2

STEROLS

Research into marine sterols falls neatly into three chronological periods. The first period encompasses the recognition in the early part of this century that cholesterol is not the sole and ubiquitous animal sterol and that marine invertebrates elaborate sterols that differ from cholesterol. The second period followed the structural elucidation of cholesterol in 1932, and is characterized by intensive investigation into the sterols of many species of marine plants and animals. This research was dominated by the pioneering efforts of the late Werner Bergmann at Yale and by several investigators in Japan, notably Tsuda's research group in Tokyo. Bergmann also faithfully and frequently reviewed marine sterol research, particularly in 1949 and (published posthumously) in 1962. The third and current period dates from the widespread availability of mass spectrometers and vapor-phase chromatographs with high resolving power. It is only now that the remarkable diversity of marine sterols, which had been foreseen by Dorée (1909) and by Bergmann (1934), is becoming an experimental reality. It is also becoming increasingly evident that few individual sterols are fully characteristic of a given plant or animal taxon. With the refinement of our analytical capabilities has come the realization that any given sterol tends to be distributed among a number of taxa, but that the sterol profile of many taxa tends to be characteristic of the taxon.

Credit for the isolation of the first marine sterol, spongosterol, from the Mediterranean sponge *Suberites domuncula* goes to Henze (1904, 1908), but it was Dorée (1909) who first surveyed a number of animal phyla for occurrence of sterols. The clear recognition that sterol diversity is greatest among the more primitive phyla was expressed by Bergmann (1949, 1962), although the full extent of this recognition was not realized until after his death because the necessary research tools were not available earlier. Bergmann furthermore possessed a great deal of intuitive know-how about sterol chemistry, and he suspected unusual structural features that he, through no fault of his, was unable to prove. In marine sterol research three independent factors have converged leaving considerable doubt about the correctness of much of the early work. First, the remarkable diversity and complexity of marine sterol mixtures. Second, the notorious ability of even complex sterol mixtures to crystallize easily and well and to melt within a narrow range. Third, the monotony of elemental composition of sterols, which has rendered virtually all marine sterol research prior to about 1960 subject to serious doubt. Notable exceptions to this generalization are the fortuitous cases in which natural sterol mixtures were either relatively simple or they contained one predominant constituent. In those cases resolution of the mixture was feasible by column chromatography, particularly when the separation was carried out via the colored 4-azobenzenecarboxylates, a method that was first introduced in 1938 (Ladenburg *et al.*), but was fashioned into a viable experimental tool by Idler and Baumann (1952). Because of these circumstances well-authenticated, largely recent work will receive major emphasis in the discussion that follows. A recent general review that includes marine sterols was published in 1970 (Brooks, 1970), but none of the more current work of great interest is covered, because of the normal time lag between manuscript and published book. In addition, sterols of invertebrates have been reviewed by Austin (1970) molluscan sterols by Idler and Wiseman (1972), echinoderm sterols by Goad and co-workers (1972), and crustacean sterols by Idler and Wiseman (1971). Much new research on marine sterols is now in progress.

Arrangement of the material according to carbon content appears to be simple and straightforward, although carbon content per se is no longer considered to be a significant property. For many years the carbon range of sterols extended from C_{27} to C_{29}, but recent discoveries of a norcholesterol and of gorgosterol and its analogs have increased the carbon range of marine sterols from C_{26} to C_{30}; yet even now C_{29} sterols predominate. It is worth noting that the carbon variation occurs exclusively in the side chain,* and not unexpectedly most of it at C-24. What structural diversity other than carbon content exists among marine sterols is almost entirely found in the side chain also, with the notable exception of the highly oxygenated C_{27} moulting hormones (Section 2, B,6) and of some bile alcohols (Section 2, B,9).

* Gibbons *et al.* (1968) in their study of the green algae *Enteromorpha intestinalis* and *Ulva lactuca* report indications of the presence of lophenol (a 4-methylsterol) derivatives. None of these has as yet been isolated.

There are a number of early reports (e.g., Bergmann *et al.*, 1945) on the isolation of marine sterols that are saturated in the nucleus, such as cholestanol and 24α-methylcholest-22-en-3β-ol (22,23-dehydrocampestanol). The latter is referred to as spongesterol, mp 150°, $[\alpha]_D +10°$, in the current literature (Brooks, 1970) and is presumably what Bergmann *et al.* (1945) called neospongosterol, when they isolated from the sponge *Suberites compacta*, a sterol that appeared to be identical with Henze's (1904, 1908) first marine sterol, spongosterol, but which they could separate into cholestanol and new sterol, mp 153°, $[\alpha]_D +10°$. A modern reinvestigation (Erdman* and Thomson, 1972) of the sterols of two sponges *Cliona celata* and *Hymeniacidon perleve* has finally confirmed the presence of cholestanol and other sterols possessing a saturated sterol nucleus. In the sponge *Hymeniacidon perleve* cholestanol is in fact the predominant sterol.

Pertinent data for the sterols are collected in Table 2.1 at the end of this chapter.

A. C$_{26}$ Sterol

An unprecedented sterol with a C-7 side chain, 22-*trans*-24-*norcholesta*-5,22-diene-3β-ol (**1**) wás isolated by Idler *et al.* (1970) from the scallop (a

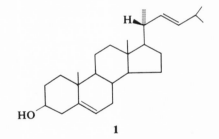

1

pelecypod mollusk, or bivalve) *Placopecten magellanicus*, where it occurs as part of a mixture of at least ten components. It was isolated by glc either as the trimethylsilyl or the methyl ether, and its structure was determined by spectral methods. The same norcholestadiene was detected by glc in other pelecypods and on the basis of retention-time comparison it is likely to occur in crustaceans and red algae as well.

Fryberg and co-workers (1971a, 1972) have synthesized the C$_{26}$ sterol via a Wittig reaction on the C-22 aldehyde. The requisite aldehyde was prepared from stigmasterol (**21**) by a new degradation sequence that involved selective bromination of the 5,6-double bond with iodobenzene dibromide (Fryberg *et al.*, 1971b). Synthetic and natural sterols were shown to be identical.

* I am indebted to Dr. Erdman for a preprint of this paper.

Three C_{26} sterols, including **1**, have been reported from the tunicate *Halocynthia roretzi* by Alcaide *et al.* (1971). The two isomers appear to have their side-chain unsaturation at C-23—one is a cholestane, the other a Δ^7-cholestene.

B. C₂₇ Sterols

If one recalls the structural parameters of the known marine sterols, which were mentioned above, one is not surprised that—except for the remarkable crustecdysone, marthasterone, and scymnol—only five conventional C_{27} sterols have so far been recorded. In addition to cholesterol, which is still the most important of all sterols, we have a Δ^7-analog of cholesterol, two compounds that differ from cholesterol by an additional double bond in the side chain, and one Δ^7-compound with an additional side-chain double bond. Discussion of the strikingly different, oxygenated steroids will follow the account of the five conventional sterols.

1. Cholesterol (Cholest-5-en-3β-ol)

Cholesterol (**2**) is hardly unique to marine organisms; nevertheless it should be included here, for it is doubtless the most commonly occurring sterol even

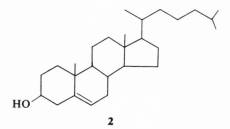

2

among marine invertebrates. Furthermore, for many years cholesterol was believed to be *the* animal sterol; this tenet was shown to be erroneous by Tsuda *et al.* (1957a, 1958a) who first demonstrated the occurrence of cholesterol in a plant—the red alga *Rhodoglossum pulcherum*. With improved separation techniques cholesterol has been shown to occur widely in marine organisms, sometimes as the predominant constituent, at times as a minor or even a trace constituent. The presence or absence of a given sterol in an organism is no longer a burning question. Of far greater interest are the questions of why a given organism will synthesize its characteristic complement of sterols and how this biosynthetic control is achieved.

2. CHOLEST-7-EN-3β-OL (Δ⁷-CHOLESTENOL, LATHOSTEROL)

Δ⁷-Cholestenol (**3**) does not appear to be a major constituent of many marine organisms that have been investigated so far. Toyama and Takagi

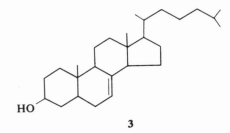

HO

3

(1953) first reported its isolation from a chiton (a mollusk of the class Amphineura), *Liolophura japonica,* and believed it to be the major sterol; but without modern tools to assess homogeneity available at that time, their statement is open to question. The same workers also reported its isolation from the sea stars *Asterias amurensis* (Toyama and Takagi, 1954) and *Asterina pectinifera* (Toyama and Takagi, 1956) and from several species of sea cucumbers (Toyama *et al.,* 1955; Takagi and Toyama, 1956). Kind and Meigs (1955) also found it to be essentially the only sterol in the mollusk *Chiton tuberculatus.*

In a careful examination of the sterols of the sea cucumber *Holothuria atra* and of the sea star *Acanthaster planci* Gupta and Scheuer (1968) found that Δ⁷-cholestenol (**3**) was the major compound among five sterols in the sea cucumber and a minor constituent (less than 10%) of the six sea-star sterols. Because Δ⁷-sterols, C_{27} to C_{30}, are a salient characteristic of the sterol mixtures of sea stars and of sea cucumbers, no doubt additional isolations will be reported.

3. CHOLESTA-5,22-DIEN-3β-OL (22-DEHYDROCHOLESTEROL)

22-Dehydrocholesterol (**4**) was prepared by partial synthesis (Bergmann and Dusza, 1958) with the suspicion that it is a naturally occurring sterol. Bergmann and Dusza (1958) based their belief on the observation that a C-22 double bond was an established feature of some C_{29} sterols. This

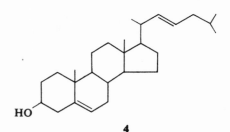

HO

4

conjecture was fully confirmed only two years later when Tsuda and co-workers (1960) isolated the sterol from the red alga *Hypnea japonica*. A further isolation was reported by the Canadian group (Tamura *et al.*, 1964) from the scallop *Placopecten magellanicus*. The Canadian workers checked the homogeneity of their sterol by glc and compared it directly with the algal compound (Tsuda *et al.*, 1960), which had been purified by the earlier technique of column chromatography of the azobenzoyl ester (Idler and Baumann, 1952).

4. CHOLESTA-5,24-DIEN-3β-OL (24-DEHYDROCHOLESTEROL, DESMOSTEROL)

Cholesta-5,24-dien-3β-ol (**5**) was isolated from a crustacean, the barnacle *Balanus glandula* by Fagerlund and Idler (1957), who confirmed its structure by partial synthesis. The compound had actually been described a year earlier under the name desmosterol from chick embryos (Stokes *et al.*, 1956), but at that time its structure had not been proved conclusively.

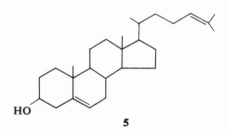

5

In recent years 24-dehydrocholesterol was shown to be the major constituent of several species of red algae (Rhodophyceae): *Porphyra purpurea* and *Rhodymenia palmata* (Gibbons *et al.*, 1967); and *Rhodymenia palmata* (dulse) and some samples of *Halosaccion ramentaceum* (Idler *et al.*, 1968). It was a minor constituent of several other species of red algae.

5. CHOLESTA-7,22-DIEN-3β-OL

The Δ⁷-analog of 22-dehydrocholesterol (**6**) has so far been a rare marine sterol. Sakai and Tsuda (1963) prepared it by partial synthesis in their

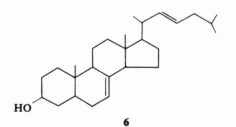

6

program of establishing a reference collection of marine sterols. Apparently only Gupta and Scheuer (1968) have identified it in natural sources as a minor sterol of an asteroid (1%) and of a holothurian (about 4%). They established the identity of this sterol by mass spectral comparison with Tsuda's synthetic sample.

6. 20-Hydroxyecdysone (Crustecdysone)

Organic chemists who attempt to isolate steroids or for that matter any given class of compounds from natural sources generally do this by a well-designed procedure that insures isolation of the sought-after substances within a narrow polar range. Such a procedure then excludes members of the class that possess unusual functionality and therefore fall outside the normal polarity range. If, on the other hand, a chemist searches for a natural product that possesses a desired physiological activity, he is guided by a bioassay and he will isolate the active substance or substances without regard to their functionality or polarity.

The account of the isolation of the principal moulting hormone of crustaceans, crustecdysone, provides a good illustration of this point. It had been known for some time that moulting of insects was governed by a hormone (ecdysone) when Karlson (1956) disclosed that an extract of the shrimp *Crangon vulgaris*, which had been prepared in analogy with insect extracts that yielded ecdysone, was active in one of the standard bioassays that employs larvae of the blowfly (*Calliphora erythrocephala*). A subsequent attempt by Karlson and Schmialek (1959) to isolate the crustacean moulting hormone from 3000 kg of shrimp was unsuccessful. Subsequently Hampshire and Horn (1966) devised a new isolation procedure, and guided by the *Calliphora* bioassay, succeeded in isolating 2 mg of "substantially pure, but noncrystalline hormone" from 1000 kg of frozen marine crayfish waste, *Jasus lalandei*. Whereas the hormone was highly active in the bioassay, it was only about one-third as active as the insect hormone ecdysone. The Australian workers (Hampshire and Horn, 1966) named the crustacean hormone crustecdysone. On the basis of spectral data and the structure of ecdysone (7) that had been

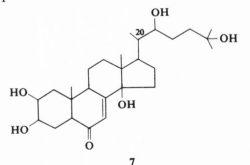

7

established by X-ray techniques (Huber and Hoppe, 1965), Hampshire and Horn (1966) proposed for crustecdysone a structure of 20-hydroxyecdysone. Horn and co-workers (1966) also succeeded in isolating crustecdysone from the pupae of an insect, the oak-silk moth *Antherea pernyi*. Shortly afterward Galbraith and Horn (1966) demonstrated the occurrence of active moulting hormones in the wood of the Australian timber tree *Podocarpus elatus*. Many other plant sources of these hormones have since been reported.

In their full paper Horn *et al.* (1968) disclosed the details of their complex isolation procedure of crustecdysone from the crayfish and cite confirmatory evidence for its structure (**8**). A synthesis of crustecdysone was reported by Hüppi and Siddall (1967).

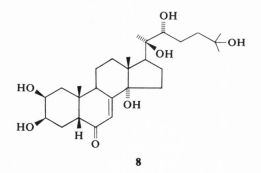

8

7. 2-DEOXYCRUSTECDYSONE

By processing 3000 kg of crayfish waste the Australian group (Galbraith *et al.*, 1968) succeeded in isolating 200 μg of a second less polar crustacean moulting hormone. From spectral data and by comparison of relative acetylation rates it was concluded that 2-deoxycrustecdysone (**9**) was the most likely structure of this hormone.

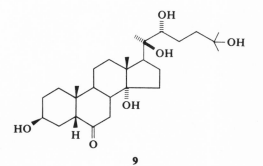

9

8. MARTHASTERONE AND DIHYDROMARTHASTERONE

A C_{27} steroid of unusual structure and unusual origin has recently been characterized. As mentioned in Chapter 1, two classes of the echinoderms, the sea cucumbers (holothurians) and the sea stars (asteroids), elaborate saponins, presumably as defensive agents. The sea cucumbers with their spectacular defense mechanism received early attention and a number of the active compounds, the holothurins, have been well characterized as a group of unique triterpenoid glycosides. The asterosaponins, on the other hand, have been studied only recently. It had been tacitly assumed that, in analogy with the holothurinogenins, the aglycones of the asterosaponins are triterpenoids. Two aglycones from the sea star *Marthasterias glacialis* have now been characterized as steroids (Turner *et al.*, 1971). Mackie and Turner (1970) earlier described the isolation of the glycoside and determined the probable sequence D-fucose–D-quinovose–D-quinovose–D-glucose–aglycone; the sulfate residue was presumably attached directly to the steroid nucleus. Hydrochloric acid hydrolysis of the glycoside led to the aglycone, which was resolvable into three components by glc. The two major constituents were separated by tlc—the uv active substance, marthasterone was assigned structure **10a** and was readily convertible into the second major component, dihydromarthasterone (**10b**) by palladium-on-charcoal hydrogenation. The structures were secured by a combination of spectral and chemical techniques.

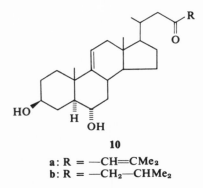

10

a: R = —CH=CMe₂
b: R = —CH₂—CHMe₂

9. SCYMNOL

A group of compounds known as the bile alcohols are a notable exception to the generalization that highly polar sterols are being overlooked unless —as was the case with the ecdysones—the isolation is monitored by bioassay. At least one of these compounds, scymnol, was first described over seventy years ago (Hammarsten, 1898) as a constituent of the bile of an Arctic shark, *Scymnus borealis*. The so-called bile salts (Haslewood, 1967) are characteristic constituents of the bile of vertebrates. In most animals these compounds are largely C-24 carboxylic acid derivatives of degraded C_{27} sterols. The derivatives generally are sodium salts of glycine (**11**) or taurine (**12**), where the amino function has formed an amide linkage with the C-24 carboxyl group. In the bile of amphibians and elasmobranch fishes (sharks and rays) the bile salts frequently are not derivatives of a C_{24} acid, but either partly or exclusively sodium sulfate esters of C_{27} alcohols.

$$H_2N—CH_2—CO_2H \qquad\qquad\qquad H_2N—CH_2—CH_2—SO_3H$$

$$\textbf{11} \qquad\qquad\qquad\qquad\qquad \textbf{12}$$

Scymnol, the first and foremost of the bile alcohols to be isolated (Hammarsten, 1898) received occasional attention over the years (Oikawa, 1925; Windaus *et al.*, 1930; Tschesche, 1931; Ashikari, 1939; Bergmann and Pace, 1943) and in time structures **13** and, somewhat later, **14** were considered plausible (Fieser and Fieser, 1959).

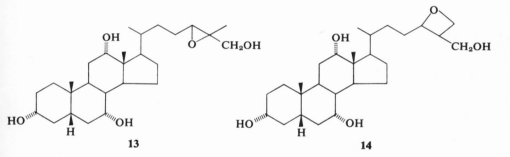

However, Haslewood in 1951 (Haslewood, 1951) first suggested that the oxide ring in scymnol (**13**) might be an artifact that arises during the customary alkaline hydrolysis of the naturally occurring sulfate ester. Cross (1960) observed that the nmr spectrum of scymnol lacked a signal for a methyl group attached to oxygen and thus concluded that structure **14** was correct. In his full paper Cross (1961) accepted Haslewood's (1951) earlier view that the scymnol of the literature was an artefact and should correctly be termed anhydroscymnol. For scymnol itself, Cross (1961) proposed structure **15**, a hexahydroxy sterol. That same year Briggs and Haslewood (1961) examined a sample of the sodium sulfate ester of scymnol from Hammarsten's (1898) original collection and arrived at the same conclusions as Cross (1961). Haslewood and co-workers (Bridgwater *et al.*, 1962) furnished the missing link that had remained: from partially acetylated scymnol sulfate by treatment in dioxane–trichloroacetic acid they isolated scymnol (**15**) itself.

Partial synthesis from a cholic acid derivative further proved that scymnol was indeed 3α,7α,12α,24ξ,26,27-hexahydroxycoprostane (15).

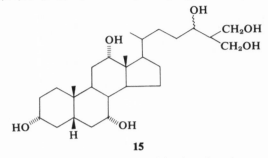

15

An interesting point in connection with the chemistry of scymnol (15) is the observation that a bile alcohol, cyprinol, isolated from the (fresh water) carp *Cyprinus carpio* is a pentahydroxycholestane of structure 16 (Hoshita *et al.*, 1963; Anderson *et al.*, 1964; Cross, 1964).

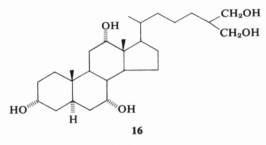

16

C. C₂₈ Sterols

All C_{28} marine sterols that have so far been characterized bear the additional carbon atom at C-24. This new center of chirality gives rise to two series of compounds epimeric with respect to C-24 in all cases where C-24 is a tetrahedral carbon. Not all possible combinations of structural features commonly identified with marine sterols (unsaturation at C-5 or C-7 combined with unsaturation at C-22 or C-24) have been reported and the configuration of the C-24 methyl group has not been ascertained in all cases. Up to the present it appears that seven C_{28} marine sterols have been well characterized.

1. 24α-METHYLCHOLEST-5-EN-3β-OL (CAMPESTEROL)

Campesterol (17) apparently has not been isolated as the major sterol component of a marine organism. It is an authentic constituent of the sterols of rape (Fernholz and Ruigh, 1941) and other terrestrial seed oils and has been reported as a possible minor sterol constituent of the green alga *Chaetomorpha crassa* (Ikekawa *et al.*, 1968) and of the coelenterate *Zoanthus con-*

fertus (Gupta and Scheuer, 1969). It has been partially synthesized by Tarzia *et al.* (1967). Since neither gas–liquid chromatography nor mass spectrometry can as yet differentiate between campesterol and its C-24 epimer, 22,23-dihydrobrassicasterol, most if not all reported trace isolations need to be confirmed.

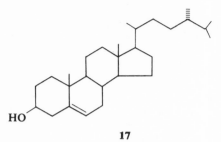

17

2. 24β-METHYLCHOLEST-5-EN-3β-OL (22,23-DIHYDROBRASSICASTEROL)

The situation with regard to 22,23-dihydrobrassicasterol (**18**) is similar to that of its epimer, campesterol (**17**). There have been several partial syntheses of this compound, the most recent one by Martinez *et al.* (1967), but only one of the reported isolations (Gupta, 1967; Gupta and Scheuer, 1969) is reliably documented. The sterol mixture of the coelenterate *Palythoa tuberculosa* consists of five components and is most likely identical with "palysterol," which Bergmann *et al.* (1951) had isolated from *P. mammilosa*. At that time "palysterol" was believed to be the C-20 epimer of γ-sitosterol. Gupta (1967) separated the mixture by glc. By mass spectrometry he determined that the principal constituent (65%) of the mixture was 24ξ-methylcholest-5-en-3β-ol. Comparison of melting points and specific rotations of the sterol, its acetate, and of the stanol with the literature data for campesterol (Fernholz and Ruigh, 1941) and for 22,23-dihydrobrassicasterol (Fernholz and Ruigh, 1940) showed conclusively that the coelenterate sterol was indeed 24β-methylcholest-5-en-3β-ol.

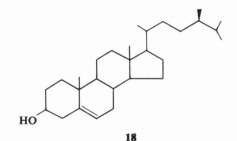

18

3. 24ξ-METHYLCHOLEST-7-EN-3β-OL (Δ⁷-ERGOSTEROL, FUNGISTEROL)

This compound, 24ξ-methylcholest-7-en-3β-ol (**19**) is the major sterol (71%) of the mixture isolated by Gupta and Scheuer (1968) from the sea star *Acanthaster planci*. It is a minor constituent (19%) in the sterol mixture of the holothurian *Holothuria atra*. These apparently constitute the only authentic isolations from marine organisms. It is not known whether the C-24 methyl group is α- or β-oriented. Knights (1967) has identified this compound as an oat seed sterol by combined gas–liquid chromatographic and mass spectrometric techniques. It had been isolated earlier (Pollock and Stevens, 1965) from ergot and other fungi.

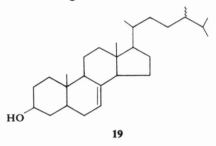

19

4. 24β-METHYLCHOLESTA-5,22-DIEN-3β-OL (BRASSICASTEROL)

Brassicasterol (**20**), generally considered a typical phytosterol, was first isolated from rapeseed oil and was well characterized. Bergmann and Ottke (1949) reported its isolation from the bivalve *Modiolus demissus*, and it is known to occur in other mollusks, e.g., the scallop *Placopecten magellanicus* (Idler *et al.*, 1964). Gupta and Scheuer (1969) identified it as the major component in the four-sterol mixture of the coelenterate *Zoanthus confertus*. It is a minor constituent of the green alga *Chaetomorpha crassa* (Ikekawa *et al.*, 1968).

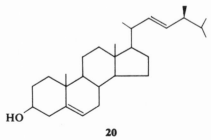

20

A supposed C_{24} epimer of brassicasterol, designated crinosterol, mp 137°–138°, $[\alpha]_D$ −47.2°, was reported by Bolker (1967). Bolker's communication was based on data of some fifteen years earlier and is therefore subject to the same doubts as are other sterol isolations of that vintage. Without adequate checks on the homogeneity (glc, mass spectrum) of Bolker's crinosterol, the authenticity of this substance should be left in abeyance.

5. 24ξ-METHYLCHOLESTA-7,22-DIEN-3β-OL (5,6-DIHYDROERGOSTEROL, α-DIHYDRO-$\Delta^{7,22}$-ERGOSTADIENOL)

The Δ^7-analog of brassicasterol with known configuration (β) at C-24 is a well-recognized minor yeast sterol (Callow, 1931; Barton and Cox, 1948). Its only reported isolation from a marine source, however with indeterminate stereochemistry at C-24 (**21**), has been reported by Gupta and Scheuer (1968) from the sea star *Acanthaster planci* and the sea cucumber *Holothuria atra*.

6. 24-METHYLCHOLESTA-5,24(28)-DIEN-3β-OL (24-METHYLENECHOLESTEROL, CHALINASTEROL, OSTREASTEROL)

24-Methylenecholesterol (**22**) is widely distributed in marine organisms and was known under a number of trivial names over the years. Idler and Fagerlund (1955) recognized it as a major sterol in oysters (*Ostrea gigas*) and clams (*Saxidomus giganteus*) and established its structure by unambiguous synthesis from the 24-oxo compound via a Wittig reaction (Idler and Fagerlund, 1957). Bergmann and Dusza (1957) paralleled this work by isolating 24-methylenecholesterol (**22**) from the coelenterate *Zoanthus proteus*, by partially synthesizing it from the 24-oxo compound; and by demonstrating its identity with chalinasterol. Gupta and Scheuer (1969) showed it to be esentially the sole sterol in a toxic coelenterate (*Palythoa* sp.) from Tahiti. It is a minor constituent of the sterol mixture of some green and brown algae (Ikekawa *et al.*, 1968). It is the second most abundant sterol (20%) in the brown algae (*Laminaria digitata* and *L. faeroensis* (Patterson, 1968). In red algae (Gibbons *et al.*, 1967; Idler *et al.*, 1968) it seems to be rare.

It should be mentioned that 24-methylenecholesterol (**22**) is not exclusively of marine origin. Knights (1967) in his definitive study of the sterol composition of oat seeds showed it to be present in the fourteen-component mixture.

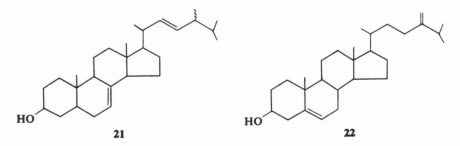

HO   **21**   HO   **22**

7. 24-METHYLCHOLESTA-7,24(28)-DIEN-3β-OL [7,24(28)-ERGOSTADIEN-3β-OL, EPISTEROL]

The Δ^7-analog of 24-methylenecholesterol (**22**) has so far been reported only once from a marine organism. Fagerlund and Idler (1959) isolated 24-methylcholesta-7,24(28)-dien-3β-ol (**23**) from asteroid *Pisaster ochraceus* by chromatography of the steryl *p*-phenylazobenzoates and secured its structure. This sterol was also shown by Knights (1967) to be present in oat seed.

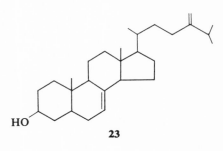

HO

23

D. C$_{29}$ Sterols

With one additional carbon atom in the side chain several new structural possibilities can arise. Not all possible structures have been found, but members of the C$_{29}$ group contain several novel and interesting features. The C-24 ethyl group that is present in all but one of the C$_{29}$ sterols may be unsaturated as an ethylidene or as a vinyl group. If it is an ethylidene group, geometrical isomerism at the 24,28-double bond may arise and one such pair (fucosterol, avenasterol) has come to light. The single example to date of a 24-vinyl group has the additional feature of a 24-hydroxy group, thus making it (saringosterol) one of the few diols among marine sterols. Another fascinating member of the C$_{29}$ group is a 23-demethylgorgosterol with the side-chain cyclopropane group.

1. 24α-ETHYLCHOLEST-5-EN-3β-OL (β-SITOSTEROL)

β-Sitosterol (**24**) is a widely distributed plant sterol, but it has long been recognized that few of the sterol samples that have been so designated over the years are pure 24α-ethylcholest-5-en-3β-ol (Brooks, 1970). Judging from a few carefully documented recent reports, β-sitosterol does not appear to be widely found in marine organisms. Its most substantial occurrence (40% of the mixture) has been reported by Ikekawa *et al.* (1968) from the green alga *Chaetomorpha crassa*. Idler *et al.* (1968) have reported small amounts in a number of red algae as have Gupta and Scheuer (1969) in the coelenterate *Palythoa tuberculosa.*

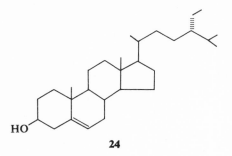

24

2. 24β-ETHYLCHOLEST-5-EN-3β-OL (CLIONASTEROL)

The 24β-epimer of β-sitosterol (**24**), on the other hand, is clionasterol (**25**) and has long been considered a marine sterol. It was probably first isolated by Dorée (1909) from the sponge *Cliona celata* and reisolated from the same animal by Valentine and Bergmann (1941). Additional structural work by Kind and Bergmann (1942) is in accord with the suggested structure. Unfortunately there had been no recent isolation to confirm homogeneity and structure. That caution needs to be exercised even when the work appears ironclad was shown by Gupta (1967) who examined the sterols of the sponge *Halichondria magnicanulosa* and showed that the mixture consisted of cholesterol, brassicasterol, and 24ξ-methylcholesterol, whereas according to Bergmann (1962) this genus of sponges is said to elaborate cholesterol or cholestanol.

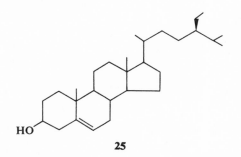

25

3. 24ξ-ETHYLCHOLEST-7-EN-3β-OL

Although the Δ⁷-analog of clionasterol (**25**) and/or β-sitosterol (**24**) has apparently not been isolated as a molecular entity, its occurrence has been demonstrated by deSouza and Nes (1968) in the blue-green alga *Phormidium luridum.* The crystalline sterol mixture isolated from the cultured algae was

investigated by glc and spectral methods and 24ξ-ethylcholest-7-en-3β-ol
(**26**) was the major constituent. Gupta and Scheuer (1968) reported this
sterol as a minor constituent of the echinoderms *Acanthaster planci* and
Holothuria atra. It was contaminated with some 24-ethylidene-Δ⁷-cholesterol.

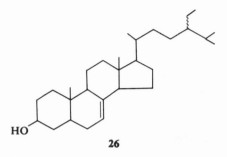

26

4. 24α-ETHYLCHOLESTA-5,22-DIEN-3β-OL (STIGMASTEROL)

Stigmasterol (**27**) is an abundant phytosterol readily available from soy-
beans. Its occurrence as a minor sterol in marine organisms has been demonstrated by Gupta and Scheuer (1968) in the brittlestar *Ophiocoma insularia*
and in the feather star *Antedon* sp. Idler *et al.* (1968) found traces of stigmasterol in some of the red algae that they examined.

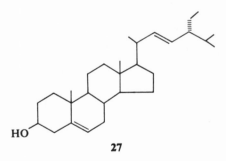

27

5. 24β-ETHYLCHOLESTA-5,22-DIEN-3β-OL (PORIFERASTEROL)

Poriferasterol (**28**) was first isolated by Valentine and Bergmann (1941)
from the sponges *Cliona celata* and *Spheciospongia vesparia* and was considered to be one of the characteristic sterols of the sponges (phylum Porifera).
Lyon and Bergmann (1942) confirmed the proposed structure by Oppenauer
oxidation to an ultraviolet active enone and by ozonization and cleavage
of the side-chain double bond. Again, no recent research has corroborated
either the homogeneity or structure of poriferasterol.

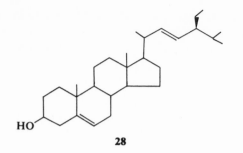

28

6. 24β-ETHYLCHOLESTA-7,22-DIEN-3β-OL (CHONDRILLASTEROL)

The Δ^7-analog of poriferasterol (**28**) is chondrillasterol (**29**) and structural definition of this sterol parallels that of its Δ^5-isomer. Bergmann and McTigue (1948) isolated the sterol from the sponge *Chondrilla nucula* and proposed its structure on the basis of degradations and comparisons that were feasible at that time. Shortly thereafter Bergmann and Feeney (1950) isolated the same sterol from the green alga *Scenedesmus obliquus D₃*. Again, recent work on this sterol is lacking.

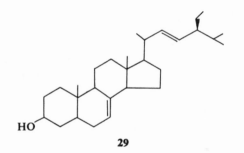

29

7. 24-ETHYLCHOLESTA-5,22(28)-DIEN-3β-OL (FUCOSTEROL)

Fucosterol (**30**) is generally considered to be the principal sterol of brown algae and represents one of the fortuitous exceptions to the general remarks

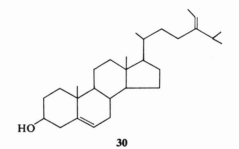

30

in the introduction to this chapter. Although fucosterol was first isolated by Heilbron *et al.* (1934) over thirty-five years ago from the brown alga *Fucus vesiculosus*, it was a homogeneous entity and its structure was determined unambiguously by degradation and partial synthesis. Nes and co-workers (1966) confirmed its distinguishing structural feature, the 24-ethylidene group, by an nmr study. Ikekawa and co-workers (1968) in their recent quantitative examination of the sterols of nine species of brown algae found that fucosterol is the major sterol in all nine species. Its percentage varied from a low of 75.5 in *Costaria costata* to a high of 97 in *Pelvetia wrightii* and *Dictyopteris divaricata*.

8. SARGASTEROL (20-ISOFUCOSTEROL)

Sargasterol (**31**) is a unique C-20 epimer of fucosterol. It was isolated by Tsuda and co-workers (1957b, 1958b) from the brown alga *Sargassum ringgoldianum*. Its structure as the C-20 epimer of fucosterol was inferred from the nonidentity of its 24-oxo product of ozonization with the corresponding degradation product of fucosterol. Hayatsu (1957) confirmed the suggested sargasterol structure by partial synthesis. In a follow-up study Tsuda *et al.* (1958c) identified the sterols in a number of brown algae by careful chromatography of the 4-azobenzene carboxylates. Only two species, *Sargassum ringgoldianum* and *Eisenia bicyclis*, contained a mixture of sargasterol and fucosterol and only in *S. ringgoldianum* was sargasterol the major sterol. All other species that were examined, including one in the genus *Sargassum*, *S. thunbergii*, were found to contain only fucosterol.

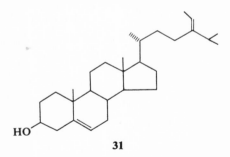

31

Interestingly, the recent paper by Ikekawa *et al.* (1968) reports the results of an investigation of the sterols of *S. ringgoldianum* by glc–mass spectrometry—fucosterol (91.5%), 24-methylenecholesterol (4.5%), saringosterol (2.1%), and cholesterol (1.7%). The authors fail to comment on the absence of sargasterol in their sample of *S. ringgoldianum*.*

---

* Gupta (1967) needed sargasterol for comparison purposes and did isolate it from *S. ringgoldianum* that had been collected in Japanese waters.

9. 28-ISOFUCOSTEROL (Δ⁵-AVENASTEROL)

Tsuda and Sakai (1960) examined the sterols of the green alga *Enteromorpha linza* and isolated an homogeneous sterol to which they assigned structure **32** on the basis of comparison of physical constants with those published for Δ⁵-avenasterol. Δ⁵-Avenasterol had been isolated by Idler *et al.* (1953) from oat seeds and was inadvertently synthesized by Dusza (1960) who attempted to synthesize the geometrical isomer fucosterol. More recently Gibbons *et al.* (1968) isolated 28-isofucosterol (**32**) as the principal sterol in the green algae *Enteromorpha intestinalis* and *Ulva lactuca*.

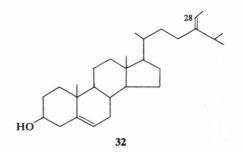

32

10. 24-ETHYLCHOLESTA-5,28(29)-DIEN-3β,24ξ-DIOL (SARINGOSTEROL)

The first dihydroxy steroid to be isolated from a marine source is saringosterol (**33**), which Ikekawa *et al.* (1966) discovered as a minor sterol in the brown algae *Sargassum ringgoldianum* and *Dictyopteris divaricata*. Its structure was determined by spectroscopic methods and by partial synthesis from 24-oxocholesterol via ethynylation and partial hydrogenation. In a footnote, the authors mention the absence of sargasterol (fucosterol was the major sterol) in *S. ringgoldianum* and suggest that it might be caused by seasonal variation. In their quantitative examination of nine species of brown algae Ikekawa *et al.* (1968) detected saringosterol varying from a trace in *Sargassum thunbergii* to 8% in *Sargassum confusum*. Patterson (1968) found saringosterol (4%) along with fucosterol (74%) and 24-methylenecholesterol (20%) in brown algae of the genus *Laminaria*.

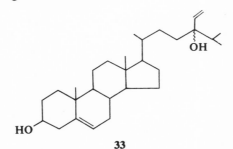

33

Saringosterol has two unusual features—a 24-vinyl (rather than the customary ethylidene) group and a 24-hydroxy group. Only scymnol (15) (see Section B,9) shares with saringosterol the 24-hydroxy function.

11. DEMETHYLGORGOSTEROL

Although this unusual sterol, which was isolated by Schmitz and Pattabhiraman (1970) from the coelenterates *Gorgonia flabellum* or *G. ventilina*, is a C_{29} sterol, it is more conveniently discussed with the C_{30} sterols, since it shares with several of these compounds the characteristic cyclopropyl group in the side chain.

E. C₃₀ Sterols

Four of these unusual C_{30} sterols have been isolated in recent years, but a fifth, demethylgorgosterol (see Section D,11), while a C_{29} compound, is really a member of this group. Unlike the C_{30} triterpenoids none of these compounds possesses an additional carbon atom in the cyclopentanoperhydrophenanthrene portion of the molecule. All additional carbon atoms are in the side chain and the characteristic feature of all but one of these compounds is the cyclopropane ring bridging carbon atoms 22 and 23. The cyclopropyl-bearing compounds will be discussed first.

1. GORGOSTEROL

Bergmann *et al.* (1943) in the course of their examination of the sterols of marine invertebrates studied the sterols of the gorgonian *Plexaura flexuosa*. Quantitative saponification of the steryl acetate indicated an unusually high molecular weight of 474 corresponding to a free sterol of composition $C_{30}H_{50}O$ to $C_{31}H_{54}O$. Further purification via the dibromoacetate led to a substance, mp 180–185°, $[\alpha]_D$ −45°, that was designated gorgosterol by Bergmann *et al.* (1943). In a review article Bergmann (1962) wrote with intuitive foresight, "Such exceptionally high melting points set these sterols apart from other levorotatory Δ^5-sterols and suggest the presence of a side chain or other structural features not heretofore encountered."

Gupta (1967) studied the sterols of some coelenterates, among them the zoanthid *Palythoa tuberculosa*. The physical constants of the sterol component were nearly identical with those reported by Bergmann *et al.* (1951) for "palysterol" isolated from *Palythoa mammilosa* and then believed to be homogeneous and isomeric with γ-sitosterol. A direct comparison of Gupta's *Palythoa* sterol and Bergmann's "palysterol" was not possible since no sample of "palysterol" could be secured. A mass spectrum of Gupta's *Palythoa* sterol showed this sterol to be a mixture of six components of molecular weights 386, 398, 400, 412, 414, and 426 (Gupta and Scheuer, 1969). By preparative glc five components could be separated (the 412/414 pair remained unresolved). The interesting constituent of molecular weight 426 had an unusually long retention time and was present in about 20% of the mixture. Its physical constants, mp 180°–182°, $[\alpha]_D$ − 44.7°, suggested that this sterol might be Bergmann's gorgosterol.

This proved to be correct when Gupta made a direct comparison with authentic gorgosterol supplied by Ciereszko (Ciereszko *et al.*, 1968). Ciereszko *et al.* (1968) had examined the sterols of a number of coelenterates, including the gorgonian *Plexaura flexuosa* (Bergmann's original source of gorgosterol), by mass spectrometry and showed the presence of a compound of molecular weight 426 in most of the animals. Ciereszko *et al.* (1968) further showed that the occurrence of gorgosterol was associated with the zooxanthellae, the symbiotic unicellular algae. Ciereszko *et al.* (1968) also made the interesting observation that under anaerobic conditions gorgosterol is reduced in the zooanthellae to a dihydrogorgosterol, $C_{30}H_{52}O$, of molecular weight 428. The existence of a dihydrogorgosterol of molecular weight 428 allows for an interesting speculation as to the identity of Attaway and Parker's (1970) sterol of molecular weight 428 that they detected in a sterol mixture derived from a recent marine sediment taken at Baffin Bay, Texas. Cholesterol (C_{27}), campesterol (C_{28}), stigmasterol (C_{29}), and β-sitosterol (C_{29}) were the identified sterols.

Gupta (1967) concluded from the signals of a 100 MHz nmr spectrum of gorgosterol that the unusual feature of gorgosterol was a cyclopropane moiety, but his attempts to open the ring either reductively or by hydrogen halide treatment failed.

In a cooperative continuation of this research Hale *et al.* (1970) succeeded in opening the cyclopropane ring of dihydrogorgosteryl acetate with hydrochloric acid in acetic acid. The ensuing olefinic mixture was ozonized and the resulting carbonyl compounds were steam-distilled into a 2,4-dinitrophenylhydrazone solution. One of the volatile components was identical with 3,4-dimethylpenta-2-one (**34**) and the nonvolatile ketone was identical with synthetic 3β-acetoxy-5α-norcholan-22-one (**35**), thus indicating that gorgo-

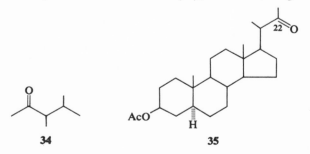

sterol has an alkyl branch at C-22. The complete structure, including absolute configuration, was secured by X-ray diffraction determination of 3-bromo-gorgostene obtained from gorgosterol by treatment with aluminum bromide (Ling *et al.*, 1970) and was shown to be (22*R*, 23*R*, 24*R*)-23,24-dimethyl-cholest-5-en-3β-ol (**36**). While alkylation at C-24 is common, the C-23 methyl group and the three-membered ring between carbon atoms 22 and 23 are without precedent.

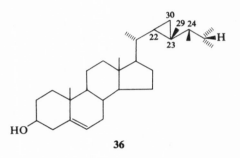

36

2. ACANTHASTEROL

Gupta (1967) in his examination of echinoderm sterols showed that the sterol mixture of the sea star *Acanthaster planci* consisted of six compounds. Furthermore one of the sterols had an unusually long glc retention time and the spectral characteristics of this sterol were strongly reminiscent of gorgosterol (Gupta and Scheuer, 1968). Gupta and Scheuer (1968) were aware of the Δ⁷ structural feature of sea star sterols and suggested therefore that this new sterol, acanthasterol, was the Δ⁷-analog of gorgosterol. Because of lack of material they were unable to prove it. Sheikh *et al.* (1971a) reisolated the compound from the same organism and showed by direct comparison of dihydroacanthasterol with dihydrogorgosterol that Gupta and Scheuer's (1968) assumption was correct and that acanthasterol had structure **37**. Sheikh *et al.* (1971a) erroneously coined for this sterol a new trivial name (acansterol), which they (Sheikh *et al.*, 1971b) subsequently withdrew.

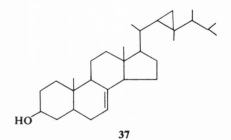

37

3. 9,11-SECOGORGOST-5-EN-3,11-DIOL-9-ONE

Perhaps the most unusual C$_{30}$ marine sterol is the gorgosterol derivative with a cleaved ring C that was isolated by Weinheimer *et al.*, (1970) from the gorgonian *Pseudopterogorgia americana* along with gorgosterol and was assigned structure **38** on the basis of chemical and spectral evidence. Interestingly, the 5α6α-epoxide was also isolated from the gorgonian. Full experimental details of these fascinating compounds are still lacking, but structure **38** was confirmed by X-ray diffraction of the 3β-*p*-iodobenzoate (Enwall *et al.*, 1972).

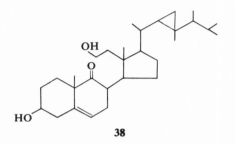

38

4. 23-DEMETHYLGORGOSTEROL

The final compound in the group, 23-demethylgorgosterol (**39**) is, as had been pointed out, a C$_{29}$ sterol, but generically it belongs with this group of C$_{30}$ compounds. As mentioned in Section D,11 it was isolated by Schmitz and Pattabhiraman (1970) from two gorgonians. Its structure was determined by spectral methods and confirmed by an X-ray diffraction study of the 3-iodobenzoate (Hsu and vander Helm, 1971; Enwall *et al.*, 1972).

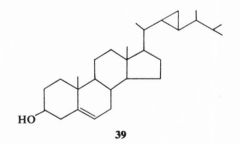

39

5. 29-METHYLISOFUCOSTEROL

A C$_{30}$ sterol without a cyclopropane in the side chain, but with a propylidene group at C-24 was recently reported by Idler and co-workers (1971, 1972). These workers isolated this sterol (**40**) from the scallop *Placopecten* *magellanicus*, where it occurs to the extent of 1.2% in a mixture of at least eleven sterols, by a combination of preparative thin-layer chromatography on silver nitrate impregnated silica gel plates and preparative glc. Its structure was secured by spectral methods. It will be recalled that the same group isolated the unusual C_{26} sterol (Section A) from the same Atlantic mollusk.

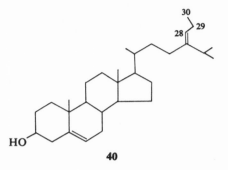

40

F. Sterol Biosynthesis

Although sterol biosynthesis has been a lively topic for research and discussion (for a pertinent review see, e.g., Clayton, 1965), the specific problem of marine sterol biosynthesis has been essentially neglected. Nomura *et al.* (1969a) carried out tracer experiments on the holothurian *Stichopus japonicus*. These investigators found that the animal did not incorporate labeled acetate into sterols or triterpenoids. Only squalene carried a label.

In a companion experiment Nomura *et al.* (1969b) investigated the unsaponifiable fraction of the holothurians *Stichopus japonicus* and *Holothuria tubulosa* and detected Δ^5- and Δ^7-sterols with 27, 28, and 29 carbon atoms. More interestingly from a biogenetic standpoint, the authors report the presence of the triterpenoids cycloartenol and lanosterol in the sea cucumbers.

In a related observation Gibbons *et al.* (1968) report the possible occurrence of cycloartenol, 24-methylenecycloartenol as well as 24-ethylidene and 24-methylenelophenol in the green algae *Enteromorpha intestinalis* and *Ulva lactuca*.

In contrast with the negative incorporation results of Nomura *et al.* (1969a), Fagerlund and Idler (1960) were able to isolate radioactive digitonin-precipitable material, presumably sterols, from the mussel *Mytilus californianus* and the clam *Saxidomus giganteus* after feeding experiments with acetate labeled with carbon-14 at C-2.

In an unrelated experiment the same researchers (Fagerlund and Idler, 1960) showed by feeding radiocholesterol ($4\text{-}^{14}C$) to the sea star *Pisaster ochraceus* that this echinoderm is capable of transforming a Δ^5- to a Δ^7-sterol.

The biosynthetic steps of sterol side-chain modification are actively being investigated. In a recent study, for example, Knapp and co-workers (1971) showed that the phytoflagellate *Ochromonas malhamensis* converts 24-ethylidene sterols, e.g., fucosterol (**30**) into poriferasterol (**28**), which possesses a 24-ethyl group. The intermediate detailed steps remain to be determined, but Knapp *et al.* (1971) showed that the efficiency of the conversion depended on the geometry of the ethylidene group.

TABLE 2.1

MARINE STEROLS

Text no.	Name (composition)	mp (in degrees)	$[\alpha]_D^\circ$	Reference
1	22-*trans*-24-Norcholesta-5,22-dien-3β-ol ($C_{26}H_{42}O$)	138–140	−52	Idler *et al.* (1970)
2	Cholesterol ($C_{27}H_{46}O$)	147–148	−40	Tsuda *et al.* (1957a, 1958a)
3	Δ⁷-Cholestenol ($C_{27}H_{46}O$)	122–122.5	+4.3 0	Toyama and Takagi (1953); Kind and Meigs (1955)
4	22-Dehydrocholesterol ($C_{27}H_{44}O$)	134.5–135.5	−56.9	Tsuda *et al.* (1960); Tamura *et al.* (1964)
5	24-Dehydrocholesterol ($C_{27}H_{44}O$)	117	−38.7	Fagerlund and Idler (1957)
6	Δ⁷,²²-Cholestadien-3β-ol ($C_{27}H_{44}O$)	123–124	−13.3	Gupta and Scheuer (1968); Sakai and Tsuda (1963)
8	Crustecdysone ($C_{27}H_{44}O_7$)	243	+61.8	Jizba *et al.* (1967)
9	2-Deoxycrustecdysone ($C_{27}H_{44}O_6$)	No data	No data	Galbraith *et al.* (1968)
10a	Marthasterone ($C_{27}H_{42}O_3$)	—	—	Turner *et al.* (1971)
10b	Dihydromarthasterone ($C_{27}H_{44}C_3$)	167–169	—	Turner *et al.* (1971)
15	Scymnol (dihydrate); 3α,7α,12α,24ξ,26,27-Hexahydroxycoprostane ($C_{27}H_{48}O_6 \cdot 2H_2O$)	120–123; 190	+34 ± 2	Bridgwater *et al.* (1962)
17	Campesterol ($C_{28}H_{48}O$)	158–159	−37	Nishioka *et al.* (1965)
18	Dihydrobrassicasterol ($C_{28}H_{48}O$)	152–153 148–149	−47 −55	Gupta (1967); Martinez *et al.* (1967)

(*continued*)

TABLE 2.1—*continued*

Text no.	Name (composition)	mp (in degrees)	$[\alpha]_D^\circ$	Reference
19	Δ^7-Ergosterol ($C_{28}H_{48}O$)	145–146 152	No rotation −20	Gupta and Scheuer (1968); *Dict. Org. Cpds.*, p. 2156.
20	Brassicasterol ($C_{28}H_{46}O$)	146–147	−58.6	Gupta and Scheuer (1969); Gupta (1967)
21	$\Delta^{7,22}$-Ergostadienol ($C_{28}H_{46}O$)	154–156 176	−20	Gupta and Scheuer (1968); Gupta (1967); Barton and Cox (1948)
22	24-Methylenecholesterol ($C_{28}H_{46}O$)	142	−35	Idler and Fagerlund (1955, 1957)
23	7,24(28)-Ergostadien-3β-ol ($C_{28}H_{46}O$)	131	+6	Fagerlund and Idler (1959)
24	β-Sitosterol; 24α-Ethyl-cholest-5-en-3β-ol ($C_{29}H_{50}O$)	137–138	−39	Nishioka *et al.* (1965)
25	Clionasterol ($C_{29}H_{50}O$)	148	−43	Kind and Bergmann (1942)
26	24ξ-Ethylcholest-7-en-3β-ol ($C_{29}H_{50}O$)	No data	No data	deSouza and Nes (1968); Gupta and Scheuer (1968)
27	Stigmasterol ($C_{29}H_{50}O$)	167–168	−50	Nishioka *et al.* (1965)
28	Poriferasterol ($C_{29}H_{48}O$)	156	−49	Valentine and Bergmann (1941)
29	Chondrillasterol ($C_{29}H_{48}O$)	169	−1	Bergmann and McTigue (1948)
30	Fucosterol ($C_{29}H_{48}O$)	124	−38.42	Heilbron, *et al.* (1934)
31	Sargasterol ($C_{29}H_{48}O$)	132–133.5	−47.5	Tsuda *et al.* (1958b)
32	Avenasterol ($C_{29}H_{48}O$)	135.5–136.4	−35.8	Tsuda and Sakai (1960)
33	Saringosterol ($C_{29}H_{48}O_2$)	160–161	−31	Ikekawa *et al.* (1966)
36	Gorgosterol	184–185	−45	Bergmann *et al.* (1943)
37	Acanthasterol (Acan-sterol) ($C_{30}H_{50}O$)	179–180	+5 ± 3	Sheikh *et al.* (1971a,b)
38	9,11-*seco*-Gorgost-5-in-3,11-diol-9-one ($C_{30}H_{50}O_3$)	No data	No data	Weinheimer *et al.* (1970)
39	23-Demethylgorgosterol ($C_{29}H_{48}O$)	162–163	−34.5	Schmitz and Pattabhiraman (1970)
40	29-Methylisofucosterol ($C_{30}H_{50}O$)	111–112	−27	Idler *et al.* (1972)

REFERENCES

Alcaide, M., Viala, J., Pinte, F., Itoh, M., Nomura, T., and Barbier, M. (1971). *C. R. Acad. Sci. C* **273**, 1386.

Anderson, I. G., Briggs, T., and Haslewood, G. A. D. (1964). *Biochem. J.* **90**, 303.

Ashikari, H. (1939). *J. Biochem. (Tokyo)* **29**, 319.

Attaway, D., and Parker, P. L. (1970). *Science* **169**, 674.

Austin, J. (1970). *In* "Advances in Steroid Biochemistry and Pharmacology" (M. H. Briggs, ed.), Vol. 1, pp. 73–96. Academic Press, New York.

Barton, D. H. R., and Cox, J. D. (1948). *J. Chem. Soc.* 1354.

Bergmann, W. (1934). *J. Biol. Chem.* **104**, 553.

Bergmann, W. (1949). *J. Mar. Res.* **8**, 137.

Bergmann, W. (1962). *In* "Comparative Biochemistry" (M. Florkin and H. S. Mason, eds.), Vol. III, Part A, pp. 103–162. Academic Press, New York.

Bergmann, W., and Dusza, J. P. (1957). *Justus Liebigs Ann. Chem.* **603**, 36.

Bergmann, W., and Dusza, J. P. (1958). *J. Org. Chem.* **23**, 1245.

Bergmann, W., and Feeney, R. J. (1950). *J. Org. Chem.* **15**, 812.

Bergmann, W., and McTigue, F. H. (1948). *J. Org. Chem.* **13**. 738.

Bergmann, W., and Ottke, R. C. (1949). *J. Org. Chem.* **14**, 1085.

Bergmann, W., and Pace, W. T. (1943). *J. Amer. Chem. Soc.* **65**, 477.

Bergmann, W., McLean, M. J., and Lester, D. J. (1943). *J. Org. Chem.* **8**, 271.

Bergmann, W., Gould, D. H., and Low, E. M. (1945). *J. Org. Chem.* **10**, 570.

Bergmann, W., Feeney, R. J., and Swift, A. N. (1951). *J. Org. Chem.* **16**, 1337.

Bolker, H. I. (1967). *Nature (London)* **213**, 905.

Bridgwater, R. J., Briggs, T., and Haslewood, G. A. D. (1962). *Biochem. J.* **82**, 285.

Briggs, T., and Haslewood, G. A. D. (1961). *Biochem. J.* **79**, 5P.

Brooks, C. J. W. (1970). *In* "Rodd's Chemistry of Carbon Compounds" (S. Coffey, ed.), 2nd ed., Vol. II, Part D, pp. 152–154, 157–159. Elsevier, Amsterdam.

Callow, R. K. (1931). *Biochem. J.* **25**, 87.

Ciereszko, L. S., Johnson, M. A., Schmidt, R. W., and Koons, C. B. (1968). *Comp. Biochem. Physiol.* **24**, 899.

Clayton, R. B. (1965). *Quart. Rev. Chem. Soc.* **19**, 168, 201.

Cross, A. D. (1960). *Proc. Chem. Soc.* 344.

Cross, A. D. (1961). *J. Chem. Soc.* 2817.

Cross, A. D. (1964). *Biochem. J.* **90**, 308.

deSouza, N. J., and Nes, W. R. (1968). *Science* **162**, 363.

Dorée, C. (1909). *Biochem. J.* **4**, 72.

Dusza, J. P. (1960). *J. Org. Chem.* **25**, 93.

Enwall, E. L., van der Helm, D., Hsu, I. N., Pattabhiraman, T., Schmitz, F. J., Spraggins, R. L., and Weinheimer, A. J. (1972). *J. Chem. Soc. Chem. Commun.* 215.

Erdman, T. R. and Thomson, R. H. (1972). *Tetrahedron* **28**, 5163.

Fagerlund, U. H. M., and Idler, D. R. (1957). *J. Amer. Chem. Soc.* **79**, 6473.

Fagerlund, U. H. M., and Idler, D. R. (1959). *J. Amer. Chem. Soc.* **81**, 401.

Fagerlund, U. H. M., and Idler, D. R. (1960). *Can. J. Biochem. Physiol.* **38**, 997.

Fernholz, E., and Ruigh, W. L. (1940). *J. Amer. Chem. Soc.* **62**, 3346.

Fernholz, E., and Ruigh, W. L. (1941). *J. Amer. Chem. Soc.* **63**, 1157.

Fieser, L. F., and Fieser, M. (1959). "Steroids," pp. 431–432. Van Nostrand-Reinhold, Princeton, New Jersey.

Fryberg, M., Oehlschlager, A. C., and Unrau, A. M. (1971a). *J. Chem. Soc. D* 1194.

Fryberg, M., Oehlschlager, A. C., and Unrau, A. M. (1971b). *Tetrahedron* **27**, 1261.

Fryberg, M., Oehlschlager, A. C., and Unrau, A. M. (1972). *J. Chem. Soc. Chem. Commun.* 204.

Galbraith, M. N., and Horn, D. H. S. (1966). *Chem. Commun.* 905.

Galbraith, M. N., Horn, D. H. S., Middleton, E. J., and Hackney, R. J. (1968). *Chem. Commun.* 83.

Gibbons, G. F., Goad, L. J., and Goodwin, T. W. (1967). *Phytochemistry* **6**, 677.

Gibbons, G. F., Goad, L. J., and Goodwin, T. W. (1968). *Phytochemistry* **7**, 983.

Goad, L. J., Rubinstein, I., and Smith, A. G. (1972). *Proc. Roy. Soc. Ser. B* **180**, 223.

Gupta, K. C. (1967). Ph.D. Dissertation, University of Hawaii, Honolulu.

Gupta, K. C., and Scheuer, P. J. (1968). *Tetrahedron* **24**, 5831.

Gupta, K. C., and Scheuer, P. J. (1969). *Steroids* **13**, 343.

Hale, R. L., Leclerq, J., Tursch, B., Djerassi, C., Gross, R. A., J., Weinheimer, A. J., Gupta, K. C., and Scheuer, P. J. (1970). *J. Amer. Chem. Soc.* **92**, 2119.

Hammarsten, O. (1898). *Hoppe-Seyler's Z. Physiol. Chem.* **24**, 322.

Hampshire, F., and Horn, D. H. S. (1966). *Chem. Commun.* 37.

Haslewood, G. A. D. (1951). *Biochem. Soc. Symp.* **6**, 83.

Haslewood, G. A. D. (1960). *Ann. N.Y. Acad. Sci.* **90**, 877.

Haslewood, G. A. D. (1967). "Bile Salts." Methuen, London.

Hayatsu, R. (1957). *Chem. Pharm. Bull.* **5**, 452.

Heilbron, I., Phipers, R. F., and Wright, H. R. (1934). *J. Chem. Soc. C* 1572.

Henze, M. (1904). *Hoppe-Seyler's Z. Physiol. Chem.* **41**, 109.

Henze, M. (1908). *Hoppe-Seyler's Z. Physiol. Chem.* **55**, 427.

Horn, D. H. S., Fabbri, S., Hampshire, F., and Lowe, M. E. (1968). *Biochem. J.* **109**, 399.

Horn, D. H., Middleton, E. J., Wunderlich, J. A., and Hampshire, F. (1966). *Chem. Commun.* 339

Hoshita, T., Nagayoshi, S., and Kazuno, T. (1963). *J. Biochem. (Tokyo)* **54**, 369.

Hsu, I-N., and van der Helm, D. (1971). *Abstracts*, Meeting of the Amer. Crystall. Assoc. p. 36.

Huber, R., and Hoppe, W. (1965). *Chem. Ber.* **98**, 2403.

Hüppi, G., and Siddall, J. B. (1967). *J. Amer. Chem. Soc.* **89**, 6790.

Idler, D. R., and Baumann, C. A. (1952). *J. Biol. Chem.* **195**, 623.

Idler, D. R., and Fagerlund, U. H. M. (1955). *J. Amer. Chem. Soc.* **77**, 4142.

Idler, D. R., and Fagerlund, D. H. M. (1957). *J. Amer. Chem. Soc.* **79**, 1988.

Idler, D. R., and Wiseman, P. (1971). *Int. J. Biochem.* **2**, 91.

Idler, D. R., and Wiseman, P. (1972). *J. Fish. Res. Bd. Can.* **29**, 385.

Idler, D. R., Nicksic, S. W., Johnson, D. R., Meloche, V. W., Schuette, H. A., and Baumann, C. A. (1953). *J. Amer. Chem. Soc.* **75**, 1712.

Idler, D. R., Tamura, T., and Wainai, T. (1964). *J. Fish. Res. Bd. Can.* **21**, 1035.

Idler, D. R., Saito, A., and Wiseman, P. (1968). *Steroids* **11**, 465.

Idler, D. R., Wiseman, P. M., and Safe, L. M. (1970). *Steroids* **16**, 451.

Idler, D. R., Safe, L., and Wiseman, P. (1971). *Abstracts.* 54th Canadian Chemical Conference, Halifax, N.S., p. 44.

Idler, D. R., Safe, L. M., and MacDonald, E. F. (1972). *Steroids* In press.

Ikekawa, N., Tsuda, K., and Morisaki, N. (1966). *Chem. Ind. (London)* 1179.

Ikekawa, N., Morisaki, N., Tsuda, K., and Yoshida, T. (1968). *Steroids* **12**, 41.

Jizba, J., Herout, V., and Sorm, F. (1967). *Tetrahedron Lett.* 1689.

Karlson, P. (1956). *Vitam. Horm. (New York)* **14**, 227.

Karlson, P., and Schmialek, P. (1959). *Hoppe-Seyler's Z. Physiol. Chem.* **316**, 83.

Kind, C. A., and Bergmann, W. (1942). *J. Org. Chem.* **7**, 341.

Kind, C. A., and Meigs, R. A. (1955). *J. Org. Chem.* **20**, 1116.

Knapp, F. F., Greig, J. B., Goad, L. J., and Goodwin, T. W. (1971). *J. Chem. Soc. D.* 707.

Knights, B. A. (1967). *Phytochemistry* **6**, 407.

Ladenburg, K., Fernholz, E., and Wallis, E. S. (1938). *J. Org. Chem.* **3**, 294.

Ling, N. C., Hale, R. L., and Djerassi, C. (1970). *J. Amer. Chem. Soc.* **92**, 5281.

Lyon, A. M., and Bergmann, W. (1942). *J. Org. Chem.* **7**, 428.

Mackie, A. M., and Turner, A. B. (1970). *Biochem. J.* **117**, 543.

Martinez, A., Romeo. A., and Tortorella, V. (1967). *Gazz. Chim. Ital.* **97**, 96.

Nes, W. R., Castle, M., McClanahan, J. L., and Settine, J. M. (1966). *Steroids* **8**, 655.

Nishioka, I., Ikekawa, N., Yagi, A., Kawasaki, T., and Tsukamoto, T. (1965). *Chem. Pharm. Bull.* **13**, 379.

Nomura, T., Tsuchiya, Y., André, D., and Barbier, M. (1969a). *Nippon Suisan Gakkaishi* **35**, 299.

Nomura, T., Tsuchiya, Y., André, D., and Barbier, M. (1969b). *Nippon Suisan Gakkaishi* **35**, 293.

Oikawa, S. (1925). *J. Biochem. (Tokyo)* **5**, 63.

Patterson, G. W. (1968). *Comp. Biochem. Physiol.* **24**, 501.

Pollock, J. R. A., and Stevens, R., Eds. (1965). "Dictionary of Organic Compounds," 4th ed. Oxford Univ. Press, New York, p. 2156.

Sakai, K., and Tsuda, K. (1963). *Chem. Pharm. Bull.* **11**, 529.

Schmitz, F. J., and Pattabhiraman, T. (1970). *J. Amer. Chem. Soc.* **92**, 6073.

Sheikh, Y. M., Djerassi, C., and Tursch, B. M. (1971a). *Chem. Commun.* 217.

Sheikh, Y. M., Djerassi, C., and Tursch, B. M. (1971b). *Chem. Commun.* 600.

Stokes, W. M., Fish, W. A., and Hickey, F. C. (1956). *J. Biol. Chem.* **220**, 415.

Takagi, T. and Toyama, Y. (1956). *Mem. Fac. Eng. Nagoya Univ.* **8**, 177.

Tamura, T., Wainai, T., Truscott, B., and Idler, D. R. (1964). *Can. J. Biochem.* **42**, 1331.

Tarzia, G., Tortorella, V., and Romeo, A. (1967). *Gazz. Chim. Ital.* **97**, 102.

Toyama, Y., and Takagi, T. (1953). *Bull. Chem. Soc. Jap.* **26**, 497.

Toyama, Y., and Takagi, T. (1954). *Bull. Chem. Soc. Jap.* **27**, 421.

Toyama, Y., and Takagi, T. (1956). *Bull. Chem. Soc. Jap.* **29**, 317.

Toyama, Y., Takagi, T., and Tanaka, Y. (1955). *Mem. Fac. Eng. Nagoya Univ.* **7**, 28.

Tschesche, R. (1931). *Hoppe-Seyler's Z. Physiol. Chem.* **203**, 263.

Tsuda, K., and Sakai, K. (1960). *Chem. Pharm. Bull.* **8**, 554.

Tsuda, K., Akagi, S., and Kishida, Y. (1957a). *Science* **126**, 927.

Tsuda, K., Akagi, S., Kishida, Y., and Hayatsu, R. (1957b). *Chem. Pharm. Bull.* **5**, 85.

Tsuda, K., Akagi, S., and Kishida, Y. (1958a). *Chem. Pharm Bull.* **6**, 101.

Tsuda, K., Hayatsu, R., Kishida, Y., and Akagi, S. (1958b). *J. Amer. Chem. Soc.* **80**, 921.

Tsuda, K., Akagi, S., Kishida, Y., Hayatsu, R., and Sakai, K. (1958c). *Chem. Pharm. Bull.* **6**, 724.

Tsuda, K., Sakai, K., Tanabe, K., and Kishida, Y. (1960). *J. Amer. Chem. Soc.* **82**, 1442.

Turner, A. B., Smith, D. S. H., and Mackie, A. M. (1971). *Nature (London)* **233**, 209.

Valentine, F. R., Jr., and Bergmann, W. (1941). *J. Org. Chem.* **6**, 452.

Weinheimer, A. J., Spraggins, R. L., and Bhacca, W. S. (1970). *Abstracts, Pacific Conference on Chemistry and Spectroscopy*, San Francisco, p. 48.

Windaus, A., Bergmann, W., and König, G. (1930). *Hoppe-Seyler's Z. Physiol. Chem.* **189**, 148.

3

BENZENOIDS

A. Simple Benzene Derivatives

1. PHENOLS AND DERIVATIVES

The observation that the red alga *Polysiphonia fastigiata* (syn. *P. lanosa*) contains a phenolic substance was perhaps first recorded by Colin and Augier (1939). In a follow-up study Leman (1944) isolated the substance as an off-white crystalline powder and confirmed its phenolic character. Results of elemental analyses further seemed to point toward the presence of two potassium sulfonate groups and one *chlorine* atom. Five years later, Mastagli and Augier (1949) reisolated this phenol from the same red alga and characterized it as a dibromo-dipotassium sulfonate-hydroxybenzoic acid, which by brief treatment with 5% hydrochloric acid on a steam bath was converted to a crystalline dibromohydroxybenzoic acid. These conditions would seem rather mild and without precedent for desulfonation of a benzenesulfonic acid. Mastagli and Augier (1949) incidentally made no reference to Leman's (1944) chlorine analysis. It may be surprising, but it is a fact, that through 1970 no organic chlorine compound had been unequivocally identified from a marine organism. Bromine, on the other hand, had long been recognized as an algal constituent, although valence state and bonding of the bromine were subjects of controversy. Augier and Henry (1950) compared the bromophenol of Mastagli and Augier (1949) with a phenolic substance that Lamure (1944) had isolated from the red algae *Vidalia volubilis* and *Halophytis*

88

pinastroides and concluded that these substances are identical and perhaps characteristic of the algal family Rhodomelaceae. Augier (1953) pursued this idea and screened twenty species belonging to eleven genera (among a total of 94 genera in the family) of the Rhodomelaceae for the presence of bromophenols by the color that an alcoholic extract produced with ferric chloride. Eight species were found to give a positive reaction and were therefore believed to contain bromophenols.

The first fully documented structure of a bromophenol was published in Japan (Saito and Ando, 1955). From *Polysiphonia morrowii* these authors isolated 5-bromo-3,4-dihydroxybenzaldehyde (**1**) and secured its identity by comparison with a synthetic sample, obtained by bromination of 3,4-dihydroxybenzaldehyde. Augier and Mastagli (1956) designated an incompletely characterized bromophenol from the red alga *Halophytis incurvus* as a dihydroxy-dipotassium sulfonate-monobromobenzoic acid. Again, as in the earlier case of the dibromophenol (Mastagli and Augier, 1949), desulfonation to a monobromodihydroxybenzoic acid of unspecified substitution pattern was said to be accomplished by brief heating on a boiling water bath with 5% aqueous hydrochloric acid.

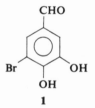

1

Hodgkin and co-workers (1966) returned to the algal species, *Polysiphonia lanosa*, on which the French workers (*vide supra*) had made their original observations and from which they had reported isolation of two partially characterized bromophenols. The Canadian workers (Hodgkin *et al.*, 1966) isolated from this alga in very high yield (1–5% of dry weight) dipotassium-2,3-dibromobenzyl alcohol-4,5-disulfate (**2**). Hydrolysis of the salt by warming it with dilute (0.1 N) hydrochloric acid yielded the corresponding 2,3-dibromo-4,5-dihydroxybenzyl alcohol (**3**), which could be reconverted to **2** by treatment in the cold with an aqueous solution of potassium carbonate

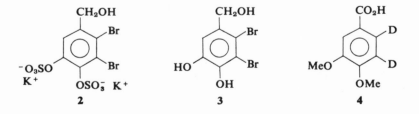

        2   **3**   **4**

and sulfur trioxide–pyridine complex. Spectral analysis of compounds **2** and **3** and transformation of **3** to 2,3-dideuterio-4,5-dimethoxybenzoic acid (**4**) and comparison with a synthetic sample secured the structure of **2**.

Katsui and co-workers (1967) isolated two phenols from the alga *Rhodomela larix*. The structure of the first, 2,3-dibromo-4,5-dihydroxybenzaldehyde (**5**) was determined on the basis of the spectral data, by transformation to the known dimethylether of **5**, and by direct comparison with an independently prepared synthetic sample.

The second compound that Katsui *et al.* (1967) isolated from this alga was assigned structure **6**. This structure, 2,3-dibromo-4,5-dihydroxybenzyl alcohol methyl ether was formulated on the basis of spectral, largely nmr, analysis of compound **6**, its 4,5-diacetate, and its 4,5-dimethyl ether. Compound **6** was also reported to be identical with an algal constituent that had been isolated by Matsumoto and Kagawa* from *Odonthalia corymbifera*

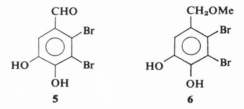

and had presumably also been synthesized. The fact that Katsui *et al.*'s isolation scheme involves the use of methanol and of acid renders the methyl ether function of **6** open to question. Furthermore, the use of acid during isolation raises the question whether **6** occurs in the plant as the free phenol or as a readily hydrolyzed ester as, e.g., **2**. In fact, it is conceivable that one of the constituents of *Rhodomela larix* is **2** rather than **6**. This question would have to be settled by careful reisolation.

Craigie and Gruenig (1967) reported **3** (2,3-dibromo-4,5-dihydroxybenzyl alcohol) as the principal phenol in the red algae *Odonthalia dentata* and *Rhodomela confervoides*. These authors deliberately treated the residue of the original ethanolic extract with acid in order to hydrolyze any sulfate esters. A

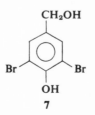

---

* In the paper by Katsui *et al.* (1967) footnote 10 refers to the work of Matsumoto and Kagawa via *Abstr. Annu. M. Chem. Soc. Jap.* **278** (1964). No further publication of this work seems to have appeared.

second phenolic substance in both algae was assigned the novel structure **7**, 3,5-dibromo-4-hydroxybenzyl alcohol, by nmr analysis of the diacetate of **7** and by direct comparison with a synthetic sample.

2. A BENZOQUINONE DERIVATIVE

A single benzoquinone has so far been reported from a marine organism. Moore and co-workers (1966a) isolated from the spines of the sea urchin *Echinothrix diadema* the unexpected 2-ethyl-3,6-dihydroxy-1,4-benzoquinone (**8**) in addition to the expected naphthoquinone derivatives. The spectral data pointed uniquely to structure **8**. Moore *et al.* (1966a) further confirmed the structure by synthesis—hydrogen peroxide oxidation of ethylhydroquinone (**9**) in base led to **8** in analogy with a procedure described for the nonalkylated compound by Jones and Shonle (1945). A second synthesis by Vadlamani

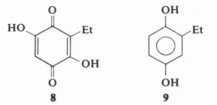

(1966) involved alkylation of 2,5-dihydroxybenzoquinone with propionyl peroxide in acetic acid patterned after earlier work by Fieser and Oxford (1942), and also led to the desired quinone **8**. Yields for both procedures were only about 25%.

B. Flavonoids

Flavonoid compounds are phenols having fifteen carbon atoms—two benzenoid nuclei connected by a three-carbon unit. One of the rings generally exhibits resorcinol (1, 3) or phloroglucinol (1, 3, 5) hydroxylation pattern, while the other is normally hydroxylated in 4; 3, 4; or 3, 4, 5 positions. The connecting unit is often cyclized. Flavonoids are widely distributed in nature and their occurrence and biosynthesis are well documented (Geissman, 1962; Geissman and Crout, 1969). In spite of this wide distribution of flavonoids there appears to be only a single well-documented isolation from marine algae (Markham and Porter, 1969). From the green alga *Nitella hookeri* these authors succeeded in separating unmistakable flavonoids. Even in this case the small amount of isolates precluded complete identification. However, the authors were able by comparison of ultraviolet spectra, chromatographic

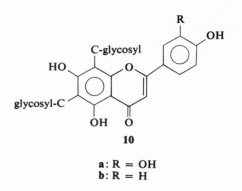

a: R = OH
b: R = H

mobility, and hydrolytic reactions to conclude that the flavonoids are *C*-glycosidic flavones related to the known lucenins (**10a**) and vicenins (**10b**). These flavonoid types had been isolated from the wood of the verbenaceous tree *Vitex lucens*, a native of New Zealand (Seikel *et al.*, 1966).

C. Naphthalene Derivatives

Almost all naphthalene derivatives that have been isolated so far from marine sources—and from terrestrial sources, for that matter—are naphthoquinones. Furthermore, all of these highly colored compounds, which are fairly common pigments in a number of plant families and in microorganisms, are being elaborated by only a single phylum of animals, the marine phylum Echinodermata. An excellent review of natural products from echinoderms has been published by Grossert (1972).

1. NAPHTHOQUINONES

The history of echinoderm pigment research is fascinating and has been reviewed twice by Thomson (1957, 1971), who has been one of the major contributors to this research. Thomson's recent revision is up-to-date and encyclopedic in scope and should be consulted for specific details on individual pigments.

a. Distribution

Among the five classes of animals in the phylum Echinodermata—echinoids (sea urchins), asteroids (sea stars), ophiuroids (brittlestars), holothurians (sea cucumbers), and crinoids (sea lilies or feather stars)—the sea urchins are perhaps the most readily accessible and certainly have been studied by a variety of biologists more extensively than have the other four classes. Not surprisingly then, for a long time quinone pigment research dealt exclusively with members of this one class. It was therefore tacitly assumed that occurrence of naphthoquinone pigments was confined to the echinoids. The first documented isolation and identification of a naphthoquinone pigment from an echinoderm that was not a sea urchin was reported from a group at Kyushu University in Japan (Mukai, 1958, 1960; Yamaguchi *et al.*, 1961). Mukai and co-workers isolated from a sea cucumber, *Polycheira rufescens*, a polyhydroxynaphthoquinone which they subsequently identified as the mono-methyl ether of a known sea urchin pigment (see Section C,1,f). At about the same time Sutherland and his group in Australia (Sutherland and Wells, 1959) reported the isolation of an anthraquinone pigment from a crinoid (see Section D). This discovery pointed to confirmation of the generally held assumption that naphthoquinone pigments are essentially an exclusive property of sea urchins. Mukai's isolation from an holothurian was considered exceptional. The known distribution of naphthoquinone pigments in echinoids has been carefully documented by Anderson *et al.* (1969).

However, Scheuer and his group (Singh *et al.*, 1967) examined the pigments of the two classes of echinoderms that had not previously been studied. They were able to show that naphthoquinone pigments do occur in the asteroids and in the ophiuroids. Interestingly, Singh *et al.* (1967) found that the pigments of the sea star *Acanthaster planci* are methyl ethers of known sea urchin pigments, whereas the range of pigments in the two species of brittlestars, *Ophiocoma erinaceus* and *O. insularia*, closely parallels that found in the sea urchins. Singh *et al.* (1967) concluded from their findings that quinone pigment distribution—polyhydroxynaphthoquinones from sea urchins and brittlestars, partial methyl ethers from sea stars and sea cucumbers—was in line with other chemical parameters, triterpenoids and sterols, that linked the two respective pairs of echinoderms. This early conclusion has since been weakened by Mathieson and Thomson's (1971) discovery of methylated pigments in sea urchins. Further work is necessary on more than a few isolated species in order to clarify these relationships.

Singh *et al.* (1967) also examined one species of crinoid in an attempt to check whether this class of animals elaborates anthraquinones exclusively. These workers isolated from *Antedon* sp. small quantities of two known sea urchin pigments in addition to large amounts of nonnaphthoquinone pigments, which they did not study. Since then Sutherland's group (Kent *et al.*, 1970; Smith and Sutherland, 1971) has discovered naphthopyrone pigments in crinoids, thus making the crinoids the most versatile of the echinoderms in pigment elaboration. Interestingly, crinoids are the earliest echinoderms to appear in the fossil record and may well be the ancestors of all echinoderms. Further work on crinoid pigments should be rewarding, but is hampered by the geographically restricted availability of this class of echinoderms.

b. The Classic Spinochromes

The unique nature of echinoderm pigments was first recognized in 1885 by MacMunn in the course of a comparative study of the blood pigments of invertebrates. He had earlier examined (MacMunn, 1883) the coloring matter in the perivisceral fluid of the sea urchins (echinoids) *Echinus esculentus* (?) and *E. sphaera* and had noted a color change that was caused by alkaline reagents. MacMunn named the pigment echinochrome. Subsequently, he (MacMunn, 1885) examined the coloring matter of the perivisceral fluid of another sea urchin, *Strongylocentrotus lividus* and extended his earlier findings. He again measured the visible spectrum of the pigment; he noted the pH-dependent color change; and he observed its ready reversible transformation between oxidized and reduced states, which led him to conclude that the physiological function of the pigment must be respiratory. Interestingly enough, Fox and Hopkins (1966) some eighty years later still raise the question of the physiological function of these pigments in the echinoderms! MacMunn, incidentally, failed to crystallize the pigment. McClendon (1912) succeeded in crystallizing echinochrome, which he had extracted from the sea urchin *Arbacia pustulosa* (syn. *A. aequituberculata, A. lixula*), by the addition of iodine in potassium iodide. McClendon's analytical data were quite respectable, but showed a small percentage of nitrogen and did not allow calculation of a reasonable empirical formula. The correct molecular formula of echinochrome, $C_{12}H_{10}O_7$, was established by Ball (1936).

The first sustained research effort in this field was initiated by Lederer and Glaser (1938). These authors reisolated echinochrome from the ovaries of the sea urchin *Arbacia aequituberculata*. They were the first to use a column of calcium carbonate in a purification step and they crystallized the pigment by vacuum sublimation. They confirmed the composition of $C_{12}H_{10}O_7$ and were the first to characterize the compound adequately, including the preparation with diazomethane of mono-, di-, and trimethyl ethers of echinochrome (Glaser and Lederer, 1939). Kuhn and Wallenfels (1939) isolated echinochrome from the ovaries of *A. pustulosa* and crystallized it from dioxane–water. On the basis of only three experiments and utilizing 286 mg of pigment —a small amount by 1939 standards—Kuhn and Wallenfels (1939) proposed **11** as the structure of echinochrome. The three experiments were the preparation of a trimethyl ether with diazomethane, reductive acetylation with zinc dust, acetic anhydride, and pyridine, which led to a leucoheptaacetate; and chromic acid oxidation, which yielded propionic acid and thus proved the nature of the side chain. An additional important structural clue was gained from a fourth experiment, which was not described in detail. Rapid zinc dust distillation at 600° furnished small amounts of naphthalene. Wallenfels and Gauhe (1943) confirmed structure **11** for echinochrome by a one-step Friedel-Crafts synthesis from 2-ethyl-1,3,4-trimethoxybenzene **(12)** and dibenzoyloxymaleic anhydride **(13)** in an aluminum chloride–sodium chloride melt in a yield of 1.5–2%, after acidification and purification via a column of calcium carbonate.

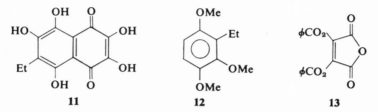

| 11 | 12 | 13 |

Lederer (1938) in the course of an extensive investigation of carotenoid pigments of invertebrates noted that a crystalline pigment extracted from the spines of *Strongylocentrotus (Paracentrotus) lividus* was not a carotenoid. Lederer and Glaser (1938) described the isolation of this pigment, which they named spinochrome, and characterized it—mp 185°, composition $C_{12}H_{10}O_8$. Accordingly, they believed it to be a hydroxyechinochrome.

Kuhn and Wallenfels (1941) isolated from the spines and tests (shells) of the sea urchin *Arbacia pustulosa* a pigment, mp 229°–230°, of composition $C_{12}H_8O_8$, which they named spinone A. In two crucial experiments, Kuhn and Wallenfels (1941) carried out a chromic acid oxidation that furnished acetic rather than propionic acid, and performed a reductive acetylation to a leucooctaacetate. On the basis of these results, and in analogy with the echinochrome structure **(11)**, these workers proposed structure **14** for spinone A. Kuhn and Wallenfels (1941) expressed the view that their spinone A **(14)**, $C_{12}H_8H_8$, might be an artifact, and that the true spine pigment might be Lederer and Glaser's (1938) spinochrome, $C_{12}H_{10}O_8$, which became oxidized during work-up, and whose structure would differ from that of spinone A by an alcoholic rather than ketonic side chain.

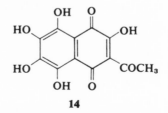

14

The distinction between echinochromes, i.e., pigments isolated from the eggs, ovaries, body fluids, etc., of echinoids (sea urchins) on one hand, and spinochromes, i.e., pigments isolated from the calcareous spines and shells of echinoids, is no longer justified. Echinochrome **(11)**, which subsequently was referred to as echinochrome A in order to distinguish it from partially characterized companion pigments, has frequently been encountered as one of the pigments occurring in the spines of the animals and, chemically, the entire group of echinoid pigments are polyhydroxynaphthoquinones (Anderson *et al.*, 1969). An earlier nomenclature proposal (Goodwin *et al.*, 1951) upheld the dual generic names and used suffixes to distinguish among individual

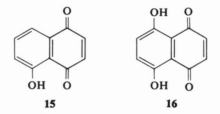

15   **16**

compounds. That proposal suffered even then from having neglected a body of research that was carried out by Kuroda and her group in Japan; however, several key structural determinations in subsequent years made the proposal obsolete. It now appears (Moore *et al.*, 1966a) that a semisystematic nomenclature based on the trivial names juglone (**15**) and naphthazarin (**16**) best serves this group of compounds. Thomson (1971) has adopted it for the second edition of his monograph.

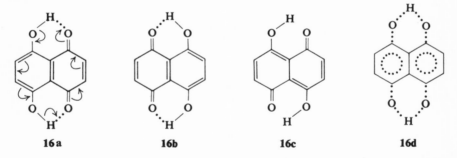

16a   **16b**   **16c**   **16d**

At the conclusion of this first period of echinoderm pigment research, dating from 1883 (MacMunn) to about 1950 (Goodwin *et al.*, 1951), one structure, that of 6-ethyl-2,3,7-trihydroxynaphthazarin (echinochrome A, **11**), had been firmly established, and another, that of 3-acetyl-2,6,7-trihydroxynaphthazarin (spinone A, **14**), was well substantiated. An ever-increasing number of additional pigments—a few echinochromes and many more spinochromes, with a bewildering alphabet soup of suffixes—was making its way into the literature. The second period of this research, dating from about 1950 to the mid-1960's was characterized by a gradual emergence of relatively few widely distributed compounds, thus displacing the earlier confusion caused by partially characterized and often poorly purified materials.

Smith and Thomson (1960, 1961) purified a sample of spinochrome E that Yoshida (1959) had isolated from the sea urchin *Psammechinus miliaris* and by the established key transformations, reductive acetylation and methyl ether derivatization, characterized the compound correctly as 2,3,6,7-tetrahydroxynaphthazarin (**17**). Confirmation by synthesis came several years later (*vide infra*).

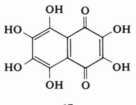

17

It is interesting to point out that these three correctly formulated pigments (echinochrome A, **11**; spinone A, **14**; and spinochrome E, **17**) have two structural features in common. All three compounds are derivatives of the symmetrical naphthazarin (**16**) and all three compounds possess fully substituted naphthalene nuclei. The fourth pigment, however, which was correctly formulated and whose structure was confirmed by synthesis during this period of research, presents an intrinsically more difficult structural problem since the pigment is a derivative of the unsymmetrical juglone (**15**).

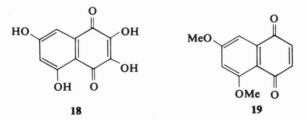

18   **19**

Spinochrome N (2,3,7-trihydroxyjuglone, **18**) was isolated by Kuroda's group in Japan from the spines of the sea urchins *Hemicentrotus pulcherrimus* and *Anthocidaris crassispina* and, on the basis of degradative work, was correctly formulated as **18** (Okajima, 1959). Smith and Thomson (1961) synthesized the pigment by hydroxylating 5,7-dimethoxy-1,4-naphthoquinone (**19**) in the 2,3 positions. Methylation with diazomethane yielded spinochrome N tetramethyl ether, from which the free pigment was obtained by demethylation with aluminum chloride–sodium chloride. Identity of the synthetic product with Kuroda's spinochrome N was determined by comparison of the published infrared and electronic spectra.

The first major break in unraveling the tangled yarn of spinochrome research came through the work of Gough and Sutherland (1964). The

Australian workers isolated a mixture of pigments from the spines and tests of the echinoid *Salmacis sphaeroides*, and by comparison with literature data surmised that one of their pigments (which was not adsorbed on calcium carbonate) might be identical with a substance referred to as spinochrome B and for which several incorrect structures had been proposed. With the aid of nmr data Gough and Sutherland (1964) proved their pigment to have the structure 2,3,7-trihydroxyjuglone (**18**), a structure that had been assigned to spinochrome N by Kuroda and that had been proved by synthesis (Smith and Thomson, 1961). By direct comparison of all available samples Gough and Sutherland (1964) showed that their pigment **18** was identical with no fewer than six other sea urchin pigments that had been described in the literature under various names, and some of them with insufficient characterization.

A decisive turning point in sea urchin pigment research came about by Chang's (1964a) use of nmr techniques and by his systematic and successful search for an adequate purification procedure. The traditional chromatography introduced by Lederer and Glaser (1938), was carried out on calcium (occasionally magnesium) carbonate. This adsorbent parallels the natural substrate of the pigments but generally does not allow elution by solvent to take place. Instead, partially developed columns are cut, redissolved, and the process is repeated. Chang (1964a) after an extensive study found that silica gel that was deactivated by prior acid treatment permitted separation of the pigments and elution by solvents.

With the aid of this tool, Chang *et al.* (1964b) isolated as the principal pigment from the spines of the sea urchins *Echinometra oblonga* and *Colobocentrotus atratus*, a compound of mp 192°–193° to which they assigned the structure 3-acetyl-2,7-dihydroxynaphthazarin (**20**) on the basis of the following evidence. Spectral data for the pigment were equally compatible with

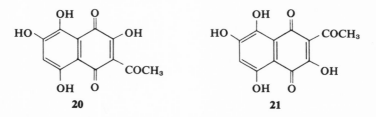

structure **20** or with its isomer, 2-acetyl-3,7-dihydroxynaphthazarin (**21**). Proof that **20** was the correct structure of this pigment was derived by treating **20** repeatedly with hot methanolic hydrogen chloride, which resulted in loss of the acetyl group (perhaps via initial hemiketal formation) and in methylation of the β-hydroxyls. Neither of the possible dimethoxynaphthazarins, **22** (2,7-) or **23** (2,6-) that could have resulted from the reaction was known at the time. Spectral data of the degradation product **22** and an unambiguous synthesis of **23** demonstrated that **20** is indeed the correct structure of the spinochrome. Not surprisingly this pigment had also been described earlier from Europe and from Japan and Chang *et al.* (1964b,c) proved its identity with spinochrome M by direct comparison and with spinochrome A by reisolation from the original source.

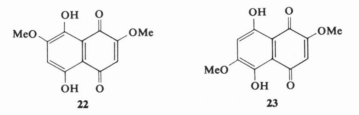

To a minor pigment isolated from the sea urchin *Echinometra oblonga* Chang *et al.* (1964c) were able to assign a structure of 3-acetyl-2,6,7-tri-hydroxynaphthazarin, largely on spectral evidence. Furthermore Chang *et al.* (1964c) showed that his pigment was identical with the correctly deduced structure **14** of spinone A and with the presumed unique and different and previously described compounds spinochrome C and spinochrome F.

By the end of 1964 a wealth of literature data on sea urchin pigments had been reduced to five authentic compounds, one juglone (**18**) and four naph-thazarin (**11, 14, 17, 20**) derivatives. The following year (Anderson *et al.*, 1965) a sixth pigment, again a naphthazarin derivative, was shown to be 2,3,6-trihydroxynaphthazarin (**24**) by synthesis via the difficultly accessible tetralone **25** and its autoxidation in the presence of potassium *t*-butoxide to **26**. The synthetic pigment **24** was shown to be identical with a pigment, spinochrome D, that Kuroda and Ohshima (1940) had isolated from the sea urchin *Pseudocentrotus depressus*.

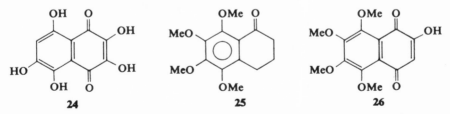

The same year, 1965, Singh *et al.* (1965) reported syntheses of spinochromes D (**24**), A (**20**), C (**14**), and E (**17**), which were remarkable for the fact that three of the compounds (**24, 14,** and **17**) resulted from a single initial conden-sation, the Friedel-Crafts reaction of chloromaleic anhydride (**27**) and 1,2-dihydroxy-3,4-dimethoxybenzene (**28**) in an aluminum chloride–sodium chloride melt. Separation of the condensation product mixture after diazo-methane treatment led to 2,3-dimethoxy-6-chloronaphthazarin (**29**), the

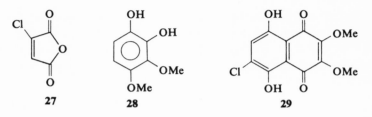

27   28   29

spinochrome D (**24**) intermediate, and 2,3-dimethoxy-6,7-dichloronaph-
thazarin (**30**), the intermediate for spinochrome E (**17**). Moreover the leuco-
acetate of spinochrome D (**31**) was acetylated with acetic acid and boron
trifluoride, furnishing spinochrome C (**14**) after hydrolysis in the presence of
air.

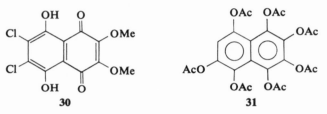

30   31

Finally, spinochrome A (**20**) was synthesized from methoxynaphthazarin
(**32**), as outlined in Scheme 3.1, via the diquinone **33**, 7-methoxy-2-hydroxy-
naphthazarin (**34**), 1,2,4,5,8-pentaacetoxy-7-methoxynaphthalene (**35**), and
2,7-diacetyl-3,6-dihydroxynaphthazarin (**36**). Removal of one of the acetyl
groups in **36** furnished **20**. Details of these syntheses were published in a full
paper (Singh *et al.*, 1968a).

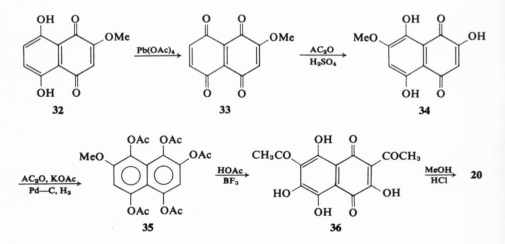

SCHEME 3.1

The following year Anderson and Thomson (1966) reported a synthesis of spinochrome E (**17**) by a different route. They followed an idea of Weygand's (1942) and condensed 3,4,5,6-tetramethoxyphthalaldehyde (**37**) with glyoxal (**38**) in the presence of potassium cyanide, a reaction that yielded directly the tetramethyl ether **39**, which after hydrogen bromide treatment led to spinochrome E (**17**).

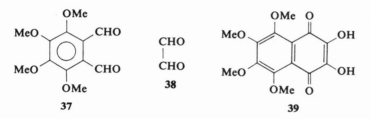

These successful, if often laborious, syntheses concluded a distinct phase of the naphthoquinone pigment research. At the end of this period it appeared that a relatively few (six) naphthoquinone derivatives accounted for what had earlier seemed to be a confusing array of closely related compounds, thus providing a simple solution to what had once seemed to be a complex problem. With the exception of Gough and Sutherland (1964), who mention a mixture of six pigments (of which they isolated one at the time), most workers reported mixtures of three or fewer compounds. It was a distinct suprise, then, when Moore and co-workers (1966a) examined the calcareous portions of two sea urchin species, *Echinothrix diadema* and *E. calamaris* and found that these animals elaborate some thirty chromatographically distinct pigments, most of which, but not all, overlap between the two species. The abundance of all but two of the pigments was very low, and while Moore *et al.* (1966a) processed 20 kg of spines, most of the pigments were available to them in quantities of 20 mg or less. In addition to the benzoquinone already discussed, this work led to the characterization of about a dozen sea urchin pigments that had not previously been encountered. Since then Gough and Sutherland (1967) have described a new juglone and Mathieson and Thomson (1971) have isolated the same juglone and two new naphthazarins as well as two unusual dimeric naphthazarins. Since much pertinent naphthoquinone chemistry has already been detailed during the preceding

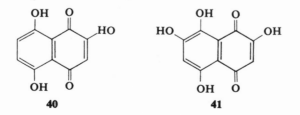

discussion of the six classic echinoid pigments, the compounds that have been characterized in this most recent phase of this research, since the mid-1960's, will be described more briefly whenever a new compound presents only a minor structural variation. Characterization of new compounds in this group was aided greatly by a series of studies that were carried out by Scheuer and his collaborators (Moore and Scheuer, 1966; Moore *et al.*, 1966b; Becher *et al.*, 1966; Piette *et al.*, 1967; Moore *et al.*, 1967; Singh *et al.* 1968b). These studies were designed to correlate various physical and chemical parameters that had become useful in the course of the spinochrome research.

c. The Naphthazarins

Naphthazarin (**16**) has in recent years been the subject of a great deal of structural scrutiny. The molecule is generally represented as 5,8-dihydroxy-1,4-naphthoquinone **16a** or **16b**.* The tautomer **16c** (4,8-dihydroxy-1,5-naphthalenedione), however, appeared to be the correct representation in the crystalline state (Pascard-Billy, 1962). As a result of more refined three-dimensional data (Cradwick and Hall, 1971) dissymmetry of the phenolic hydrogen atoms is indicated and a representation such as **16a** is favored. In solution, naphthazarin is probably best formulated as the tautomers **16a,b** and not by the fully delocalized structure **16d** since *inter alia* a dipole moment of 1.1 D is reported for naphthazarin in dilute benzene solution (Dumas *et al.*, 1969). A rapid tautomeric equilibrium (**16a,b**) is supported by the fact that at temperatures from ambient to $-60°$ only a single nmr signal is observed for the nuclear protons (Chang, 1964a; Brockmann and Zeeck, 1968).

For substituted naphthazarins in solution Moore and Scheuer (1966) have shown that the nature of the substituents determines which of the rings will have predominant quinoidal character. An X-ray crystallographic study (Fehlmann and Niggli, 1965) of the plant pigment cordeauxiaquinone (2-acetyl-7-hydroxy-6-methoxy-3-methylnaphthazarin) is in accord with the X-ray work on naphthazarin (**16**), which showed that **16c** is the predominant tautomer in the crystalline state.

The naphthazarins (**16**) because of their symmetry present fewer structural problems than do the juglones (**15**). Of the six classic echinoid pigments five are derivatives of naphthazarin—trihydroxy- (**24**), tetrahydroxy- (**17**), a monoacetyldihydroxy- (**20**), monoacetyltrihydroxy- (**14**), and ethyltrihydroxy- (**11**) naphthazarin. It is not surprising then that among the newly isolated echinoid pigments we find monohydroxynaphthazarin (naphthopurpurin), **40** and 2,7-dihydroxy- naphthazarin, (**41**). Naphthopurpurin (**40**) had long been known as a synthetic compound, while 2,7-dihydroxynaphthazarin (**41**) had been described under the trivial name mompain following its isolation from the microorganism *Helicobasidium mompa* (Nishikawa, 1962; Natori *et al.*,

---

* For structures **16a–d**, see page 96.

1965). Both of these hydroxynaphthazarins were isolated by Moore *et al.* (1966a) from *Echinothrix diadema* and *E. calamaris*.

Two further naphthazarin derivatives, isomeric monohydroxymonoacetyl compounds were also isolated from the spines of the same animals by Moore *et al.* (1966a). One of them, 2-hydroxy-3-acetylnaphthazarin, **42**, had been characterized by Moore *et al.* (1966b) as one of the numerous sodium borohydride reduction products of 3-acetyl-2,7-dihydroxynaphthazarin. The structure of the other compound, 2-hydroxy-6-acetylnaphthazarin (**43**) was elucidated by spectral data and confirmed by synthesis (Moore *et al.*, 1966a).

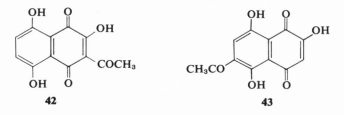

The remaining two naphthazarins that Moore *et al.* (1966a) had isolated from the two prolific *Echinothrix* species are hydroxyethylnaphthazarins. The simpler of the two, having the same substitution pattern as **43**, is 2-hydroxy-6-ethylnaphthazarin (**44**). Pigment **44**, like **42**, had been encountered by Moore *et al.* (1966b) as one of the sodium borohydride reduction products of spinochrome A (**14**). Its isomer, 2-hydroxy-3-ethylnaphthazarin (**45**), has not so far been isolated from sea urchins. Singh *et al.* (1967) isolated **45** from the calcareous portions of the brittlestars *Ophiocoma erinaceus* and *O. insularia*. Its structure was shown to be identical with one of the sodium borohydride reduction products of spinochrome A (**14**) (Moore *et al.*, 1966b). This was also true of the remaining naphthazarin derivative from *Echinothrix*, 2,7-dihydroxy-3-ethylnaphthazarin (**46**).

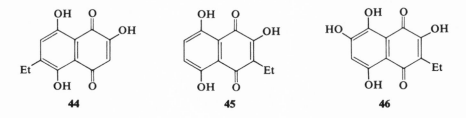

d. A Pyranonaphthazarin

In addition to the hydroxynaphthazarins, with or without two-carbon side chains, which have been described, *Echinothrix diadema* and *E. calamaris* elaborate a pigment bearing a four-carbon alkyl side chain. Moore *et al.*

(1966a) separated the pigment from the normal naphthazarins and juglones by column chromatography on deactivated silica gel. Purification was achieved by tlc and crystallization. Moore *et al.* (1968) determined its

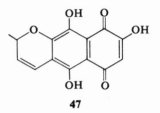

47

structure as 2-methyl-8-hydroxy-2*H*-pyrano[3,2-*g*]naphthazarin (**47**) by spectral determination and by conversion to the 3,4-dihydro compound and comparison of its spectral characteristics with those of the parent compound (**47**). Compound **47** clearly is a member of the spinochrome group of compounds, a 6-alkenyl derivative of 2,7-dihydroxynaphthazarin. Its novel feature is the uncommon four-carbon side chain that has cyclized to form an hetero ring.

e. The Dimeric Naphthazarins

Mathieson and Thomson (1971) isolated from the shells of the sea urchin *Spatangus purpureus* by cellulose column chromatography, in addition to two common naphthazarins (**11** and **24**), the first two dimeric naphthazarins. The major pigment was shown to have structure **48**, ethylidene-3,3'-bis(2,6,7-trihydroxynaphthazarin) on the basis of spectral data, conversion to its methyl ether, and by synthesis from trihydroxynaphthazarin **24**, occurring in the same animal (*vide supra*) and acetaldehyde. It is worth noting that in the mass spectrometer only the methyl ether furnishes a weak molecular ion peak. The highest mass peak (*m/e* 238) in the mass spectrum of the dimer is that of trihydroxynaphthazarin.

A second dimeric naphthazarin, also isolated by Mathieson and Thomson (1971) from the same animal, but in very much lower yield (3 mg *versus* 420 mg of **48** from 1.5 kg of starting material) has been tentatively assigned structure **49**, the anhydrocompound of **48**. The anhydro compound did

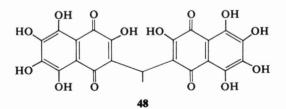

48

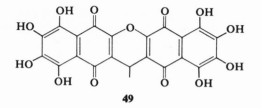

49

furnish a parent peak (*m/e* 484), which gave rise to the base peak by loss of a methyl group. Mathieson and Thomson (1971) have encountered compound **49** in two other sea urchin species.

f. Naphthazarin Methyl Ethers

Only six methyl ethers of naphthazarin derivatives have so far been isolated from echinoderms. All six are derivatives of three of the classic sea urchin pigments.

The first of this group of compounds (see part a) was isolated by Mukai (1958) and given the trivial name namakochrome. From 28,000 sea cucumbers of the species *Polycheira rufescens* Mukai isolated no less than 2.36 g of the new pigment that he correctly assumed to be a polyhydroxynaphthoquinone. Mukai (1960) subsequently deduced its correct structure as the monomethyl-ether of tetrahydroxynaphthazarin (**50**). Final proof of the structure was accomplished by its hydrolysis to spinochrome E (**17**) (Yamaguchi *et al.*, 1961).

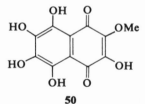

50

Two isomeric dimethyl ethers of the same spinochrome E (**17**) were isolated by Singh *et al.* (1967) from the sea star *Acanthaster planci*. Structures **51**, 2,6-dihydroxy-3,7-dimethoxynaphthazarin, and **52**. 2,7-dihydroxy-3,6-dimethoxynaphthazarin, were secured by spectral data (Moore *et al.*, 1967) and by conversion to the known tetrahydroxy- and tetramethoxynaphthazarins.

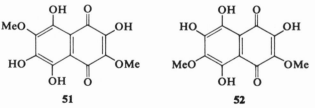

51   **52**

A monomethyl ether of spinochrome A (**20**) was recognized by Singh *et al.* (1967) as one of perhaps a dozen pigments in the brittlestars *Ophiocoma erinaceus* and *O. insularia* and its structure (**53**), 2-hydroxy-3-acetyl-7-methoxynaphthazarin, was secured by comparison with known spinochrome A derivatives (Moore *et al.*, 1967).

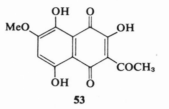

53

Finally, Mathieson and Thomson (1971) isolated two isomeric monomethyl ethers of echinochrome A (**11**) from the spines of the sea urchin *Diadema antillarum*. The structures were shown to be **54**, 2-methoxy-3,7-dihydroxy-6-ethylnaphthazarin, and **55**, 3-methoxy-2,7-dihydroxy-6-ethylnaphthazarin, by spectral data and by comparison with characteristics of these compounds that had been prepared earlier by Moore and co-workers (1967).

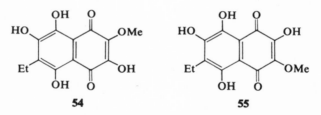

54   **55**

It is worthwhile to bear in mind the Mathieson and Thomson (1971) *caveat* regarding the isolation of their two methyl ethers **54** and **55** against the use of methylation procedures as an aid in the separation of polyhydroxynaphthoquinone pigments, since those concerned with isolation of natural products must always guard against isolating artifacts. Mathieson and Thomson's specific doubt about the structures of Moore *et al.* (1966a) that have been determined in connection with methylation procedures is probably unfounded. If the *Echinothrix* of Moore *et al.* had contained any natural methyl ethers, these compounds would have appeared in the fast-moving fractions, for which methylation was not necessary, and which were separated directly.

g. The Juglones

Only one of the six classic echinoid pigments (see Section C,1,b) is a juglone, viz. 2,3,7-trihydroxyjuglone (**18**, spinochrome B or N). It has been pointed out earlier that in contrast to naphthazarin, the unsymmetrical juglone structure gives rise to a larger number of possible isomers. Furthermore, the unsubstituted α-position in a naphthoquinone system makes these compounds more susceptible to further oxidation than naphthazarins lacking this structural feature. Perhaps because of these inherent difficulties, perhaps for other reasons not yet apparent, only six additional juglone derivatives have been reported so far in the literature. The simplest of these, 2-hydroxy-6-ethyljuglone (**56**) was isolated by Moore *et al.* (1966a) from the spines of the sea urchin *Echinothrix calamaris* in a yield of $10^{-5}\%$ (from 700 g of spines). Fortunately these workers had a sample of this compound prepared earlier by sodium borohydride reduction of spinochrome A (**20**) and so were able to characterize the new pigment by electronic spectral data and thin-layer mobility and color.

Moore *et al.* (1966a) isolated a second 6-ethyljuglone as the principal pigment in the spines of *Echinothrix diadema*—the new compound was shown to be 2,3,7-trihydroxy-6-ethyljuglone (**57**) on the basis of its spectral characteristics and those of its trimethyl ether (Moore and Scheuer, 1966). Mathieson and Thomson (1971) have since found the same pigments in the sea urchin *Temnopleurus toreumaticus*.

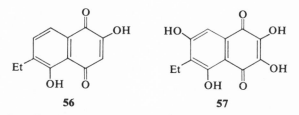

        56   **57**

An isomeric trihydroxyethyljuglone was isolated by Singh *et al.* (1967) from two brittlestars, *Ophiocoma erinaceus* and *O. insularia*. Its structure was deduced to be 2,6,7-trihydroxy-3-ethyljuglone (**58**) by spectral data, by its conversion to the known trimethyl ether (Moore *et al.*, 1967), and by comparison with a sodium borohydride reduction product of the corresponding 3-acetyl compound (Singh *et al.*, 1967).

This 2,6,7-trihydroxy-3-acetyljuglone (**59**) is a naturally occurring pigment that was first isolated using preparative paper chromatography by Gough and Sutherland (1967) from the sea urchin *Salmacis sphaeroides* and subsequently

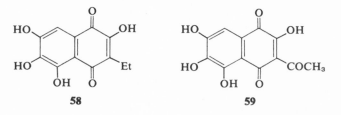

        58   **59**

by Mathieson and Thomson (1971) from *Temnopleurus toreumaticus*. Gough and Sutherland proved the structure of this juglone by using the deacetylation reaction that had been introduced by Chang *et al.* (1964b) and by converting the resulting trihydroxyjuglone to the more tractable 2,5,6,7-tetramethoxy-1, 4-naphthoquinone (**60**), which was identical with a synthetic sample (Natori and Kumada, 1965).

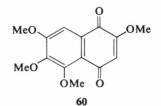

60

2,7-Dihydroxy-6-acetyljuglone (**61**) was one of the pigments isolated by Moore *et al.* (1966a) from *Echinothrix diadema* and *E. calamaris*. It could be identified with one of the known sodium borohydride reduction products of 2,7-dihydroxy-3-acetylnaphthazarin (spinochrome A, **20** Moore *et al.*, 1966b) and has since been synthesized (Singh *et al.*, 1969).

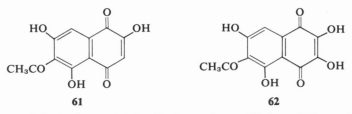

61   **62**

The last of the juglones, 2,3,7-trihydroxy-6-acetyljuglone (**62**) was isolated by the same workers (Moore *et al.*, 1966a) from the same sea urchin spines in a yield of $5 \times 10^{-5}\%$. Although the pigment itself proved to be difficult to handle (perhaps it had not been fully purified), the suggested structure **62** appeared to be the most likely one on the basis of the uv and nmr spectra of the stable 2,3-dimethyl ether. This dimethyl ether was subsequently synthesized by Singh *et al.* (1969) and was identical with the naturally derived one.

Gough and Sutherland (1970) describe a synthesis of compound **62** starting with 2,3,7-trihydroxyjuglone (spinochrome B, **18**) and arrive at a product whose melting point, color, and one electronic spectral band differ from those of the described pigment. Because the derived 2,3-dimethyl ethers of the two products proved identical and natural material was no longer available, the discrepancy remains. It may well be due to the fact that the original observations prior to methylation (Moore *et al.*, 1966a) were made on an impure pigment. Gough and Sutherland's (1970) suggestion that the original pigment is the 3-monomethyl ether of **62** is unsupported by the molecular ion in the mass spectrum.

h. Biosynthesis

In contrast to the many experiments that have been carried out over the years toward the elucidation of the biosynthetic pathways occurring in plants and microorganisms, very little work has been reported on the biosynthetic pathways that take place in marine invertebrates. Only a single paper (Salaque *et al.*, 1967) has so far dealt with the biosynthesis of naphthoquinone pigments in echinoderms. In parallel experiments sea urchins (*Arbacia pustulosa*) were fed labeled methionine, acetic, and propionic acids. Whereas incorporation of the label was poor in all cases, it was clearly best with acetic acid. Labeled echinochrome A (6-ethyl-2,3,7-trihydroxynaphthazarin, **11**) was isolated and the side chain was degraded. The principal results showed that echinoids can biosynthesize naphthoquinone pigments *de novo* and that the extremely poor incorporation of the label in the side chain may indicate that biosynthesis takes place stepwise, addition of a side chain following synthesis of the nucleus.

Oxygenation patterns and the prevalance of no or a two-carbon side chain in these pigments (Singh *et al.*, 1967) have pointed to an acetate-polyketide pathway. Further experiments will be necessary to confirm the findings of Salaque *et al.* (1967) and to fill in the details of the biosynthetic pathways.

2. NAPHTHOPYRONES

The only nonquinoid naphthalene derivatives, although no doubt biogenetically related to the quinones, to be isolated from a marine source have recently been described by Sutherland's group in Australia. From a crinoid (feather star) *Comantheria perplexa* Kent *et al.* (1970) isolated two related pigments, of which the principal component was shown to be the sodium sulfate ester of 8-hydroxy-5,6-dimethoxy-2-methyl-4*H*-naphtho-[2,3-*b*]pyran-4-one (comantherin), **63a**. Acid hydrolysis yielded the free phenol, comantherin, **63b**, whose structure was unequivocally established by spectral data and by direct comparison of comantherin methyl ether (**63c**) with the known dimethyl ether of the mold metabolite norrubrofusarin (**64**).

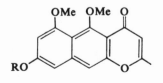

63

a: R = SO_3^- Na^+
b: R = H
c: R = Me

From the mother liquors of comantherin sulfate (**63a**) Kent *et al.* (1970) succeeded in isolating a second pigment (**65a**), which on acid hydrolysis furnished neocomantherin, (**65b**), readily shown to differ from comantherin (**63b**) only by an *n*-propyl side chain at carbon 2. As Kent *et al.* (1970) point out these naphthopyrones are almost certainly of acetate polyketide origin and represent no radical departure from other echinoderm pigments.

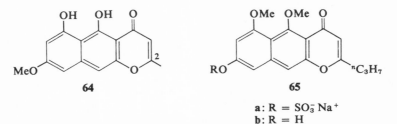

64

65

a: R = SO_3^- Na$^+$
b: R = H

From another crinoid, *Comanthus parvicirrus timorensis*, Smith and Sutherland (1971) have isolated the first angular naphthopyrones, based on a 4*H*-naphtho[1,2-*b*]pyran (**66**) skeleton. The naturally occurring pigments are three related sulfate esters, which on mild acid hydrolysis yield corresponding

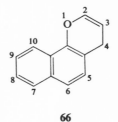

66

phenols, the structures of which were deduced largely by spectral methods. The trivial name comaparvin had been assigned to the structurally simplest pigment, 5,8-dihydroxy-10-methoxy-2-*n*-propyl-4*H*-naphtho[1,2-*b*]pyran-4-one (**67**); the other two pigments are 6-methoxycomaparvin (**68**) and 6-methoxycomaparvin-5-methyl ether (**69**). The hydroxyl group at C-8 appears to bear the sulfate group. Interestingly, a green and a yellow color variant of

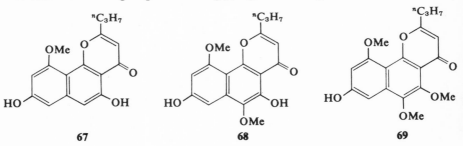

67

68

69

this crinoid species were studied and both yielded the same pigments in slightly different proportions.

Isolation from the same animal species of a 1,2-dihydroxy compound, its monomethyl, and its dimethyl ether is reminiscent of plant constituents and does not resemble what is normally found in animal pigments. Powell and Sutherland (1967a) in connection with the crinoidal anthraquinone pigments have called attention to the similarity between the pigments of lichens and crinoids, which in a way reiterates the outward plantlike appearance of many stalked crinoids (sea lilies).

D. Anthracene Derivatives

Only seven derivatives of anthracene have so far been reported from marine sources, six of them derivatives of 9,10-anthraquinone (**70**) and one a rare 1,2-anthraquinone (**71**). All six 9,10-anthraquinone (**70**) derivatives have been isolated from crinoids (feather stars, sea lilies) and have been studied by Sutherland and his collaborators in Australia—an ideal locale for crinoid research since a majority of these echinoderms have been reported from the Indopacific. The sole representative of 1,2-anthraquinone (**71**) has recently been reported from a marine annelid.

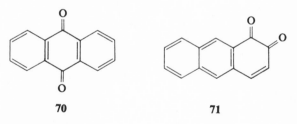

70   **71**

1. 9,10-ANTHRAQUINONES

In contrast to the large research effort on sea urchin pigments dating back over many years, the pigments of the often spectacularly beautiful crinoids have received only recent attention. The brief chemical history has been carefully detailed by Sutherland and Wells (1967). Aside from work on a fossil crinoid by Blumer (1965) who formulated the pigments as polyhydroxy-naphthodianthrones, most of the other work prior to Sutherland's did not progress beyond a few electronic spectral data.

Sutherland and Wells (1959) first reported on their crinoid research in a preliminary communication. In the full paper Sutherland and Wells (1967) described the isolation of three anthraquinone derivatives from the crinoids *Comatula pectinata* and *C. cratera*. By acetone extraction of the animals, followed by chromatography on magnesium carbonate, three crystalline pigments were obtained. By spectral analysis and chemical degradation the major pigment was shown to be 1,3-dihydroxy-6,8-dimethoxy-4-butyryl-9, 10-anthraquinone (rhodocomatulin 6,8-dimethyl ether, **72**). Hydrolysis in hydrobromic-acetic acid yielded butyric acid and a tetrahydroxyquinone,

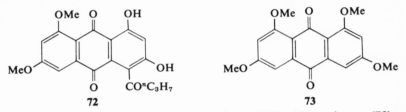

whose tetramethyl ether, 1,3,6,8-tetramethoxy-9,10-anthraquinone (**73**) was synthesized and compared (Low *et al.*, 1965). In the course of their work Sutherland and Wells (1967) developed a useful mild degradation, which eliminates the side chain in alkaline hydrosulfite ($Na_2S_2O_4$) without cleaving methyl ethers.

The second closely related pigment was shown to have structure **74**, rhodocomatulin 6-methyl ether.

The third major quinone pigment of the crinoids *Comatula pectinata* and *C. cratera* proved to have its carbon side chain at C-8 of the anthraquinone nucleus. Powell and Sutherland (1967a) named the pigment rubrocomatulin monomethyl ether and proved its structure to be 1,4,5,7-tetrahydroxy-2-methoxy-8-butyryl-9,10-anthraquinone, **75**.

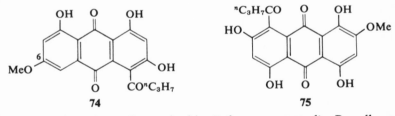

From another Australian crinoid, *Ptilometra australis* Powell *et al.* (1967b) isolated three related pigments whose structures could be secured by spectral and chemical methods. Rhodoptilometrin was shown to be 1,6,8-trihydroxy-3-(1-hydroxypropyl)-9,10-anthraquinone (**76**) while isorhodoptilometrin (**77**) differed from **72** only by the position of the side-chain

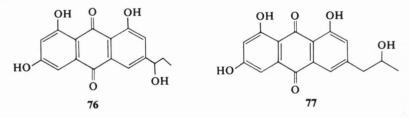

hydroxyl group. The trimethyl ether of isorhodoptilometrin (**77**) could be compared directly with the dimethyl ether of a mold metabolite nalgiovensin (Birch and Massy-Westropp, 1957).

The third of the *Ptilometra* pigments proved to be 1,6,8-trihydroxy-3-propyl-9,10-anthraquinone-2-carboxylic acid (**78**), which was given the trivial name ptilometric acid. Pyrolysis of the acid yielded carbon dioxide and the trimethyl ether of decarboxylated product was a known (Birch and Baye, 1961) substance.

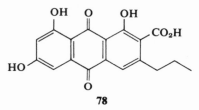

78

As a result of this work by the Australian group it has become clear that the crinoid pigments merit further attention. So far, certainly, the crinoids have shown themselves to be the most versatile of the echinoderms, having yielded anthraquinones, naphthopyrones (Kent *et al.*, 1970; Smith and Sutherland, 1971) and naphthazarins (Singh *et al.*, 1967).

2. A 1,2-ANTHRAQUINONE

A recent preliminary report (Prota *et al.*, 1971) ascribes to the red pigment hallachrome of the marine annelid *Halla parthenopeia* the structure 7-hydroxy-8-methoxy-5-methyl-1,2-anthraquinone (**79**). Hallachrome was isolated by plunging the live worms into chloroform and chromatographing the concentrated red chloroform extract on polyamide. The pigment crystallized readily from the eluate of the red band. The tentatively assigned structure **79** was ascertained largely by spectral analysis, by conversion to a crystalline derivative with *o*-phenylenediamine, and by conversion to a leucotriacetate.*

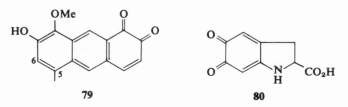

79   **80**

Two points are worth mentioning in connection with the hallachrome isolation. Hallachrome appears to be the first example of a naturally occur-

* *Note added in proof:* The full paper (Prota *et al.*, 1972) places the methyl group at C-6 rather than at C-5 as shown in **79**.

ring 1,2-anthraquinone. Its structural elucidation has had a long history (see Prota *et al.*, 1971 for references); hallachrome earlier was believed to be 2,3-dihydroindole-5,6-quinone-2-carboxylic acid (**80**) (Mazza and Stolfi, 1930). The correctness of this structure was being questioned over the years. More recent (Bielig and Möllinger, 1960) analytical data pointed to a $C_{21}H_{25}NO_9S$ formulation.*

Secondly, this work represents one of the few studies that deal with constituents of marine annelids. According to Prota *et al.* (1971) *Halla parthenopeia* is a rare polychaete found chiefly in the Bay of Naples, but in general members of the annelid class Polychaeta (the marine worms) are very common. According to Barnes (1968) these worms tend to be overlooked because of their secretive habits. Yet many are said to be strikingly beautiful and are colored red, pink, or green; and some are iridescent.

E. Miscellaneous Benzenoids

1. A BENZOPYRENE DERIVATIVE

Interestingly enough, another novel quinone pigment has recently been reported from a marine annelid, the lugworm *Arenicola marina* (Morimoto *et al.*, 1970). The pigment arenicochrome had been isolated earlier (van Duijn *et al.*, 1951; van Duijn, 1952a,b) as a tripotassium salt that liberated 2 moles of sulfuric acid on acid hydrolysis as well as a dark purple pigment, arenicochromine. Zinc dust distillation furnished benzo[*a*]pyrene (3,4-benzpyrene) (**81**). Spectral data of arenicochromine and several derivatives led Morimoto *et al.* (1970) to suggest 2,5,10-trihydroxy-4-methoxybenzo[*a*]-pyrene-6,12-quinone (**82**) as the most likely structure for arenicochromine. The positions of the sulfate esters have not yet been determined.

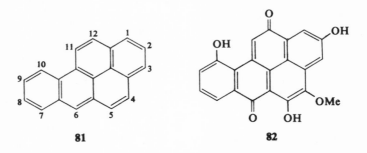

81   **82**

* The analytical samples contained appreciable (about 6%) inorganic residue. The sulfur content may indicate the possibility that hallachrome may occur as a sulfate ester, for which ample precedent exists among marine natural products. The nitrogen content remains without ready rationale.

2. A CHROMANOL

An interesting and so far unique relative of δ-tocopherol (83), which is a constituent of soybean oil and belongs to the vitamin E group of compounds, has been isolated from the brown alga *Taonia atomaria* by González and co-workers (1971). It was named taondiol and its structure (84) was deduced by spectral determinations and several chemical transformations. The unique feature of taondiol (84) is the cyclized sesquiterpene side chain, which is acyclic in the tocopherols. Table 3.1 lists the characterized marine benzenoids.

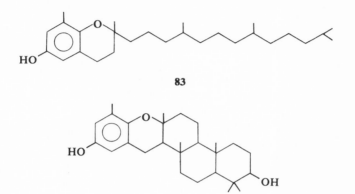

83

84

TABLE 3.1

MARINE BENZENOIDS[a]

Text no.	Name	mp (in degrees)	Reference
1	5-Bromo-3,4-dihydroxybenzaldehyde	227–228	Saito and Ando (1955)
2	Dipotassium-2,3-dibromobenzyl alcohol-4,5-disulfate	—	Hodgkin *et al.* (1966)
5	2,3-Dibromo-4,5-dihydroxy-benzaldehyde	203–205	Katsui *et al.* (1967)
6	2,3-Dibromo-4,5-dihydroxy-benzyl alcohol methyl ether	129–130	Katsui *et al.* (1967)
7	3,5-Dibromo-4-hydroxybenzyl alcohol	113–113.5	Craigie and Gruenig (1967)
8	2-Ethyl-3,6-dihydroxy-1,4-benzoquinone	130–145 (subl.)	Moore *et al.* (1966a)
11	6-Ethyl-2,3,7-trihydroxy-naphthazarin (echinochrome A)	220.5–221	Ball (1936)
14	3-Acetyl-2,6,7-trihydroxy-naphthazarin (spinone A, spinochrome C)	229–230	Kuhn and Wallenfels (1941)

(continued)

TABLE 3.1—*continued*

Text no.	Name	mp (in degrees)	Reference
17	2,3,6,7-Tetrahydroxynaphthazarin (spinochrome E)	> 350	Smith and Thomson (1960)
18	2,3,7-Trihydroxyjuglone (spinochrome N, B)	> 260 (dec)	Kuroda and Iwakura (1942)
20	3-Acetyl-2,7-dihydroxynaphthazarin (spinochrome A)	192–193	Chang *et al.* (1964b)
24	2,3,6-Trihydroxynaphthazarin (spinochrome D)	285–290 (subl.)	Kuroda and Ohshima (1940)
40	2-Hydroxynaphthazarin (naphthopurpurin)	200–210 (subl.)	Moore *et al.* (1966a)
41	2,7-Dihydroxynaphthazarin (mompain)	265–275 (subl.)	Moore *et al.* (1966a)
42	2-Hydroxy-3-acetylnaphthazarin	163–164 (dec)	Moore *et al.* (1966a)
43	2-Hydroxy-6-acetylnaphthazarin	179–180 (dec)	Moore *et al.* (1966a)
44	2-Hydroxy-6-ethylnaphthazarin	204–204.5	Moore *et al.* (1966a)
45	2-Hydroxy-3-ethylnaphthazarin	185–186	Singh *et al.* (1967)
46	2,7-Dihydroxy-3-ethylnaphthazarin	190–192	Moore *et al.* (1966a)
47	2-Methyl-8-hydroxy-2H-pyrano-[3,2-*g*]naphthazarin	165–172	Moore *et al.* (1968)
48	Ethylidene-3,3′-bis(2,6,7-tri-hydroxynaphthazarin)	155–157	Mathieson and Thomson (1971)
49	Anhydroethylidene-3,3′-bis(2,6,7-trihydroxynaphthazarin)	253–256	Mathieson and Thomson (1971)
50	2-Methoxy-3,6,7-trihydroxy-naphthazarin (namakochrome)	218	Mukai (1958)
51	2,6-Dihydroxy-3,7-dimethoxy-naphthazarin	252–254 (dec)	Singh *et al.* (1967)
52	2,7-Dihydroxy-3,6-dimethoxy-naphthazarin	218–219	Singh *et al.* (1967)
53	2-Hydroxy-3-acetyl-7-methoxy-naphthazarin	246–248	Singh *et al.* (1967)
54	2-Methoxy-3,7-dihydroxy-6-ethylnaphthazarin	202–204	Mathieson and Thomson (1971)
55	3-Methoxy-2,7-dihydroxy-6-ethylnaphthazarin	179–184	Mathieson and Thomson (1971)
56	2-Hydroxy-6-ethyljuglone	219–220	Moore *et al.* (1966a)
57	2,3,7-Trihydroxy-6-ethyljuglone	265–269 (dec, subl.)	Moore *et al.* (1966a)
58	2,6,7-Trihydroxy-3-ethyljuglone	220–226	Singh *et al.* (1967)
59	2,6,7-Trihydroxy-3-acetyljuglone	275–280 (dec)	Gough and Sutherland (1967)

(*continued*)

TABLE 3.1—*continued*

Text no.	Name	mp (in degrees)	Reference
61	2,7-Dihydroxy-6-acetyljuglone	215 (dec)	Moore *et al.* (1966a)
62	2,3,7-Trihydroxy-6-acetyljuglone	245–255 (subl.)	Moore *et al.* (1966a)
63a	8-Hydroxy-5,6-dimethoxy-2-methyl-4*H*-naphtho[2,3-*b*]pyran-4-one (comantherin) sodium sulfate	> 200 (dec)	Kent *et al.* (1970)
67	5,8-Dihydroxy-10-methoxy-2-*n*-propyl-4*H*-naphtho[1,2-b]pyran-4-one (comaparvin)	232–233 (dec)	Smith and Sutherland (1971)
68	6-Methoxycomaparvin	200–201.5 (dec)	Smith and Sutherland (1971)
69	6-Methoxycomaparvin 5-methyl ether	221–222 (dec)	Smith and Sutherland (1971)
72	1,3-Dihydroxy-6,8-dimethoxy-4-butyryl-9,10-anthraquinone (rhodocomatulin 6,8-dimethyl ether)	208.5–209 229.5–230.5	Sutherland and Wells (1967)
74	1,3,8-Trihydroxy-6-methoxy-4-butyryl-9,10-anthraquinone (rhodocomatulin 6-methyl ether)	250–252 (dec)	Sutherland and Wells (1967)
75	1,4,5,7-Tetrahydroxy-2-methoxy-8-butyryl-9,10-anthraquinone (rubrocomatulin 2-methyl ether)	298–299	Sutherland and Wells (1967)
76	1,6,8-Trihydroxy-3-(1-hydroxy-propyl)-9,10-anthraquinone (rhodoptilometrin)	217–218	Powell and Sutherland (1967b)
77	1,6,8-Trihydroxy-3-(2-hydroxy-propyl)9,10-anthraquinone (isorhodoptilometrin)	275–277	Powell and Sutherland (1967b)
78	1,6,8-Trihydroxy-3-propyl-9,10-anthraquinone-2-carboxylic acid (ptilometric acid)	298–299	Powell and Sutherland (1967b)
79	7-Hydroxy-8-methoxy-6-methyl-1,2-anthraquinone (hallachrome)	224–226 (dec)	Prota *et al.* (1971, 1972)
82	2,5,10-Trihydroxy-4-methoxy-benzo[*a*]pyrene-6,12-quinone (arenicochromine)	300 (dec)	van Duijn (1952b)
84	Taondiol	283–284	González *et al.* (1971)

a Subl (sublimation) implies volatilization or resolidification *without* structural change; dec (decomposition) implies structural change.

REFERENCES

Anderson, H. A., and Thomson, R. H. (1966). *J. Chem. Soc. C* 426.
Anderson, H. A., Smith, J., and Thomson, R. H. (1965). *J. Chem. Soc.* 2141.
Anderson, H. A., Mathieson, J. W., and Thomson, R. H. (1969). *Comp. Biochem. Physiol.* **28**, 333.
Augier, J. (1953). *Rév. Gen. Bot.* **60**, 257.
Augier, J., and Henry, M. H. (1950). *Bull. Soc. Bot. France* **97**, 29.
Augier, J., and Mastagli, P. (1956). *C. R. Acad. Sci. C* **242**, 190.
Ball, E. G. (1936). *J. Biol. Chem.* **114**, vi.
Barnes, R. D. (1968). "Invertebrate Zoology." 2nd ed. Saunders, Philadelphia, Pennsylvania.
Becher, D., Djerassi, C., Moore, R. E., Singh, H., and Scheuer, P. J. (1966). *J. Org. Chem.* **31**, 3650.
Bielig, H.-J., and Möllinger, H. (1960). *Hoppe-Seyler's Z. Physiol. Chem.* **321**, 276.
Birch, A. J., and Baye, C. J. (1961). *J. Chem. Soc. C* 4691.
Birch, A. J., and Massy-Westropp, R. A. (1957). *J. Chem. Soc.* 2215.
Blumer, M. (1965). *Science* **149**, 722 and earlier papers cited therein.
Brockmann, H., and Zeeck, A. (1968). *Chem. Ber.* **101**, 4221.
Chang, C. W. J. (1964a). Ph.D. Dissertation, University of Hawaii, Honolulu.
Chang, C. W. J., Moore, R. E., and Scheuer, P. J. (1964b). *J. Amer. Chem. Soc.* **86**, 2959.
Chang, C. W. J., Moore, R. E., and Scheuer, P. J. (1964c). *Tetrahedron Lett.* 3557.
Colin, H., and Augier, J. (1939). *C. R. Acad. Sci. C* **208**, 1450.
Cradwick, P. D., and Hall, D. (1971). *Acta. Cryst.* **B27**, 1990.
Craigie, J. S., and Gruenig, D. E. (1967). *Science* **157**, 1058.
Dumas, J.-M., Cohen, A., and Gomel, M. (1969). *C. R. Acad. Sci. C* **269**, 69.
Fehlmann, M., and Niggli, A. (1965). *Helv. Chim. Acta* **48**, 305.
Fieser, L. F., and Oxford, A. E. (1942). *J. Amer. Chem. Soc.* **64**, 2060.
Fox, D. L. and Hopkins, T. S. (1966). *In* "Physiology of Echinodermata" (R. A. Boolootian, Ed.), Chap. 12. Wiley (Interscience), New York.
Geissman, T. A., Ed. (1962). "The Chemistry of Flavonoid Compounds." Pergamon, Oxford.
Geissman, T. A., and Crout, D. H. G. (1969). "Organic Chemistry of Secondary Plant Metabolism," Chapter VII. Freeman, San Francisco, California.
Glaser, R., and Lederer, E. (1939). *C. R. Acad. Sci. C* **208**, 1939.
González, A. G., Darias, J., and Martin, J. D. (1971). *Tetrahedron Lett.* 2729.
Goodwin, T. W., Lederer, E., and Musajo, L. (1951). *Experientia* **7**, 375.
Gough, J., and Sutherland, M. D. (1964). *Tetrahedron Lett.* 269.
Gough, J. H., and Sutherland, M. D. (1967). *Aust. J. Chem.* **20**, 1693.
Gough, J. H., and Sutherland, M. D. (1970). *Aust. J. Chem.* **23**, 1839.
Grossert, J. S. (1972). *Chem. Soc. Rev.* **1**, 1.
Hodgkin, J. H., Craigie, J. S., and McInnes, A. G. (1966). *Can. J. Chem.* **44**, 74.
Jones, R. G., and Shonle, H. A. (1945). *J. Amer. Chem. Soc.* **67**, 1034.
Katsui, N., Suzuki, Y., Katamura, S., and Irie, T. (1967). *Tetrahedron* **23**, 1185.
Kent, R. A., Smith, I. R., and Sutherland, M. D. (1970). *Aust. J. Chem.* **23**, 2325.
Kuhn, R., and Wallenfels, K. (1939). *Chem. Ber.* **72**, 1407.
Kuhn, R., and Wallenfels, K. (1941). *Chem. Ber.* **74**, 1594.
Kuroda, C., and Ohshima H. (1940). *Proc. Imp. Acad. (Tokyo)* **16**, 214.
Kuroda, C., and Iwakura, H. (1942). *Proc. Imp. Acad. (Tokyo)* **18**, 74.

Lamure, J. (1944). *C. R. Acad. Sci. C* **218**, 246.

Lederer, E. (1938). *Bull. Soc. Chim. Biol.* **20**, 567.

Lederer, E., and Glaser, R. (1938). *C. R. Acad. Sci. C* **207**, 454.

Leman, A. (1944). *Bull. Soc. Chim. Fr.* **11**, 564.

Low, T. F., Park, R. J., Sutherland, M. D., and Vessey, I. (1965). *Aust. J. Chem.* **18**, 182.

MacMunn, C. A. (1883). *Proc. Birmingham Phil. Soc.* **3**, 380.

MacMunn, C. A. (1885). *Quart. J. Microsc. Sci.* **25**, 469.

Markham, K. R., and Porter, L. J. (1969). *Phytochemistry* **8**, 1777.

Mastagli, P., and Augier, J. (1949). *C. R. Acad. Sci. C* **229**, 775.

Mathieson, J. W., and Thomson, R. H. (1971). *J. Chem. Soc. C* 153.

Mazza, F. P., and Stolfi, G. (1930). *Boll. Soc. Ital. Biol. Sper.* **5**, 1121 [*Chem. Abstr.* **25**, 3402 (1931).]

McClendon, J. F. (1912). *J. Biol. Chem.* **11**, 435.

Moore, R. E., and Scheuer, P. J. (1966). *J. Org. Chem.* **31**, 3272.

Moore, R. E., Singh, H., and Scheuer, P. J. (1966a). *J. Org. Chem.* **31**, 3645.

Moore, R. E., Singh, H., Chang, C. W. J., and Scheuer, P. J. (1966b). *J. Org. Chem.* **31**, 3638.

Moore, R. E., Singh, H., Chang, C. W. J., and Scheuer, P. J. (1967). *Tetrahedron* **23**, 3271.

Moore, R. E., Singh, H., and Scheuer, P. J. (1968). *Tetrahedron Lett.* 4581.

Morimoto, I., Shaikh, M. I. N., Thomson, R. H., and Williamson, D. G. (1970). *Chem. Commun.* 550.

Mukai, T. (1958). *Mem. Fac. Sci. Kyushu Univ. Ser. C* **3**, 29. [*Chem. Abstr.* **53**, 12504 (1959).]

Mukai, T. (1960). *Bull. Chem. Soc. Jap.* **33**, 1234.

Natori, S., Kumada, Y., and Nishikawa, H. (1965). *Chem. Pharm. Bull. (Tokyo)* **13**, 633.

Nishikawa, H. (1962). *Agr. Biol. Chem.* **26**, 696.

Okajima, M. (1959). *Sci. Pap. Inst. Phys. Chem. Res. Tokyo* **53**, 356; also for full references to the earlier papers of Kuroda and her group.

Pascard-Billy, C. (1962). *Bull. Soc. Chim. Fr.* 2282, 2293, 2299.

Piette, J. H., Okamura, M., Rabold, G. P., Ogata, R. T., Moore, R. E., and Scheuer, P. J. (1967). *J. Phys. Chem.* **71**, 29.

Powell, V. H., and Sutherland, M. D. (1967a). *Aust. J. Chem.* **20**, 541.

Powell, V. H., Sutherland, M. D., and Wells, J. W. (1967b). *Aust. J. Chem.* **20**, 535.

Prota, G., D'Agostino, M., and Misuraca, G. (1971). *Experientia* **27**, 15.

Prota, G., D'Agostino, M., and Misuraca, G. (1972). *J. Chem. Soc. Perkin. Trans.* I 1614.

Saito, T., and Ando, Y. (1955). *Nippon Kagaku Zasshi* **76**, 478. [*Chem. Abstr.* **51**, 17810 (1957).]

Salaque, A., Barbier, M., and Lederer, E. (1967). *Bull. Soc. Chim. Biol.* **49**, 841.

Seikel, M. K., Chow, J. H. S., and Feldman, L. (1966). *Phytochemistry* **5**, 439.

Singh I., Moore, R. E., Chang, C. W. J., and Scheuer, P. J. (1965). *J. Amer. Chem. Soc.* **87**, 4023.

Singh H., Moore, R. E., and Scheuer, P. J. (1967). *Experientia* **23**, 624.

Singh, H., Folk, T. L., and Scheuer, P. J. (1969). *Tetrahedron* **25**, 5301.

Singh, I., Moore, R. E., Chang, C. W. J., Ogata, R. T., and Scheuer, P. J. (1968a). *Tetrahedron* **24**, 2969.

Singh, I., Ogata, R. T., Moore, R. E., Chang, C. W. J., and Scheuer, P. J. (1968b). *Tetrahedron* **24**, 6053.

Smith, I. R., and Sutherland, M. D. (1971). *Aust. J. Chem.* **24**, 1487

Smith, J., and Thomson, R. H. (1960). *Tetrahedron Lett.* (1) 10.

Smith, J., and Thomson, R. H. (1961). *J. Chem. Soc.* 1008.

Sutherland, M. D., and Wells, J. W. (1959). *Chem. Ind. (London)* 291.

Sutherland, M. D., and Wells, J. W. (1967). *Aust. J. Chem.* **20**, 515.

Thomson, R. H. (1957). "Naturally Occurring Quinones." Butterworth, London.

Thomson, R. H. (1971). "Naturally Occurring Quinones," 2nd ed. Academic Press, New York.

Vadlamani, N. L. (1966). M.S. Thesis, University of Hawaii, Honolulu.

van Duijn, P. (1952a). *Rec. Trav. Chim. Pays-Bas* **71**, 585.

van Duijn, P. (1952b). *Rec. Trav. Chim. Pays-Bas* **71**, 595.

van Duijn, P., Havinga, E., and Lignac, G. O. E. (1951). *Experientia* **7**, 376.

Wallenfels, K., and Gauhe, A. (1943). *Chem. Ber.* **76**, 325.

Weygand, F. (1942). *Chem. Ber.* **75**, 625.

Yamaguchi, M., Mukai, T., and Tsumaki, T. (1961). *Mem. Fac. Sci. Kyushu Univ. Ser. C* **4**, 193. [*Chem. Abstr.* **58**, 4486 (1963).]

Yoshida, M. (1959). *J. Mar. Biol. Ass. U.K.* **38**, 455.

4

NITROGENOUS COMPOUNDS

A. Simple Amines, Amino Acids, and Choline Derivatives

Simple amines and amino acids are not normally included in a discussion of secondary metabolites. A brief summary account is presented here since some of these compounds may be shown to be precursors of complex marine natural products; since some of them may prove to be unique to the marine environment . . . even after a good deal more research in this field will have been done than has been reported to date; and since over the years a number of these compounds have been implicated as constituents of marine toxins.

Welsh and Prock (1958), for example, in their survey of quartenary amines in coelenterates examined the role of tetramine (tetramethylammonium) (**1**) as a paralytic agent. The authors (Welsh and Prock, 1958) observed that the paralytic action of cold aqueous extracts of the tentacles of the sea

$$Me_4N^+$$

1

anemone *Metridium dianthus* is more powerful than the isolated tetramine. They therefore concluded that this discrepancy may be caused by association of tetramine (**1**) in its natural state with a specific protein. Asano and Itoh (1960), on the other hand, have shown tetramine (**1**) to be the major toxic component of the salivary glands of the marine gastropod mollusk *Neptunea arthritica*.

The question of the uniqueness to the marine environment of some simple nitrogenous compounds may be illustrated with the betaine homarine (*N*-methylpyridinium-2-carboxylate) (**2**). Homarine was first isolated by Hoppe-Seyler (1933) from the muscle of the lobster, *Homarus americanus*. Its wide distribution among marine invertebrate phyla and its apparent absence in freshwater invertebrates have been carefully investigated by Gasteiger *et al.* (1960), thereby fostering the assumption that homarine (**2**) is uniquely found in marine organisms. Yet List (1958) demonstrated its presence in the fungus *Polyporus sulphureus*, which may be an exceptional finding or may be followed in time by other isolations from terrestrial sources.

2

1. SIMPLE AMINES

Steiner and Hartmann (1968) have carried out the most extensive survey to date of the occurrence of volatile amines in marine algae. They investigated five species of green algae (Chlorophyceae), eleven species of brown algae (Phaeophyceae), and twelve species of red algae (Rhodophyceae). They identified nine volatile amines, of which trimethylamine (NMe_3) was represented in 23 of the 28 species. This survey reported the first natural occurrence of 2-methylmercaptopropylamine (**3**). However, virtually all simple mono- and difunctional amines that have been isolated at one time or another from a marine source are also known from nonmarine plants and animals.

$$CH_3-CH-CH_2NH_2$$
$$|$$
$$SMe$$

3

2. SIMPLE AMINO ACIDS

In the early 1950's, after the discovery and development of paper chromatographic techniques there was a rash of investigations that made use of these techniques. Because of early successes with paper chromatograms of amino acids, many investigators studied amino acid distribution in a great variety of organisms. Few studies at that time, however, dealt with amino acid distribution in marine phyla. Only very recently Bergquist and Hartmann (1969) have carried out an ambitious study using two-dimensional thin-layer electrophoresis and chromatography on a marine phylum, a study of the kind that for many years was commonplace for many terrestrial phyla. Bergquist and Hartman (1969) investigated 67 species belonging to 50 genera of West Indian sponges. All species belonged to a single class of the phylum Porifera, the Demospongiae. This class comprises the largest number of species in the phylum, and all but one are marine families. The authors' purpose (Bergquist and Hartmann, 1969) was to provide a chemotaxonomic survey using free amino acids as the indicator. Because of the care and thoroughness of the work it also serves as a guide to the occurrence in this class of primitive marine animals of some thirty amino acids, some of them rare, e.g., β-aminoisobutyric acid (**4**), pipecolic (piperidine-2-carboxylic) acid (**5**), and methionine sulfoxide (**6**). It is interesting to note in this connection that a 4,5-dehydropipecolic acid (L-baikiain) has been isolated from the red alga *Corallina officinalis* (Madgwick *et al.*, 1970).

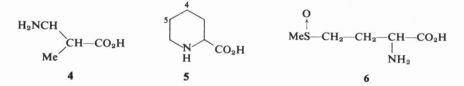

Most other studies in this field have dealt with the isolation of one or another uncommon, perhaps unique, amino acid, frequently in connection with a study of some physiological activity. An example of this type is zooanemonine (**7**), isolated by Ackermann and List (1960) from the coelenterate *Anemonia sulcata*. Other examples that may be cited are the two closely related betaines herzynine (**8**) and ergothioneine (**9**), which Ackermann and List (1958) isolated from the crustacean *Limulus polyphemus* and which previously had been known predominantly from fungi.

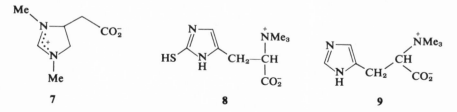

Kittredge and co-workers over the years have made valuable contributions toward the isolation and characterization of unusual amino acids in marine organisms. Kittredge *et al.* (1962) first isolated 2-aminoethylphosphonic acid (**10**) from the sea anemone *Anthopleura elegantissima* and demonstrated its uncommon carbon–phosphorus bond. In search of a simple precursor of amino acid **10** Kittredge and Hughes (1964) discovered 2-amino-3-phosphonopropionic acid (**11**) not in the same organism, but in another coelenterate, the zoanthid *Zoanthus sociatus*. Acid **11** had become available at about the same time by synthesis (Chambers and Isbell, 1964).

$$H_2NCH_2—CH_2—PO(OH)_2$$

$$HO_2C—\underset{\underset{NH_2}{|}}{CH}—CH_2—PO(OH)_2$$

10   **11**

Two *N*-methyl derivatives of acid **10**, 2-methylaminoethanephosphonic acid (**12**) and the betaine 2-trimethylammoniumethanephosphonate (**13**) were subsequently isolated and identified by Kittredge and Isbell (1967) from the sea anemone *Anthopleura xanthogrammica*. A review of the work by Kittredge and co-workers on amino acids in marine organisms is available in a U.S. Government document (Roberts and Kittredge, 1969).

$$\overset{H}{Me N}—CH_2—CH_2—PO(OH)_2$$

$$Me_3\overset{+}{N}—CH_2—CH_2—PO(O_2H)^-$$

12   **13**

Another group of uncommon amino acids, the simple aminosulfonic acids, is often associated with the study of marine organisms. Lindberg (1955a) first isolated *N*,*N*-dimethyltaurine (**14**) from the red alga *Furcellaria fastigiata* in connection with carbohydrate research. In a directed search for taurine and its derivatives in red algae Lindberg (1955b) subsequently identified dimethyltaurine (**14**) as well as *N*-methyltaurine (**15**) and taurine (**16**) itself in *Ptilota pectinata*, *Porphyra umbilicalis*, and *Gelidium cartilagineum*. Ackermann and Pant (1961) were the first to demonstrate the occurrence of the two *N*-methyltaurines (**14** and **15**) in an invertebrate, the sponge *Calyx nereis*. Ciereszko *et al.* (1960) isolated the missing methyl derivative, 2-trimethylammoniumethanesulfonate (taurobetaine, **17**) from the gorgonian *Briareum asbestinum*.

$$Me_2N—CH_2—CH_2—SO_3H$$

$$\overset{H}{Me N}—CH_2—CH_2—SO_3H$$

14   **15**

$$H_2N—CH_2—CH_2—SO_3H$$

$$Me_3\overset{+}{N}—CH_2—CH_2—SO_3^-$$

16   **17**

Two amino acids, both guanidylurea derivatives and both probably unique metabolites have been isolated and identified by Hashimoto and his group. Ito and Hashimoto (1965) first isolated one of the substances from the red alga *Gymnogongrus flabelliformis* and showed it to be γ-(guanylureido)-butyric acid (**18**) by aqueous decomposition at 120° in a sealed tube to carbon dioxide, guanidine, and γ-aminobutyric acid, by permanganate oxidation to guanylurea and succinic acid, and by barium hydroxide hydrolysis to γ-ureidobutyric acid. Subsequently, Ito and Hashimoto (1966a) assigned the trivial name gongrine to the new amino acid.

$$\underset{\textbf{18}}{H_2N-\overset{\displaystyle NH}{\overset{\|}{C}}-\overset{H}{N}-CO-\overset{H}{N}-(CH_2)_3-CO_2H}$$

The second component, gigartinine, was also isolated by Ito and Hashimoto (1966a,b) from the same red alga by ion exchange chromatography. Hydrolytic and oxidative degradations paralleled those for gongrine. Aqueous and acidic decomposition at 120° led to carbon dioxide, guanidine, and L-ornithine (**19**). Oxidative reaction yielded guanylurea and succinic acid as before, while barium hydroxide treatment furnished L-ornithine (**19**) and citrulline (**20**). These transformations pointed to α-amino-γ(guanylureido)-valeric acid (**21**) as the structure of gigartinine. Ito and Hashimoto (1969) synthesized both new amino acids by standard procedures and established their identities. Minor discrepancies in the infrared spectra (nujol mull) of natural and synthetic gigartinine were no doubt caused by the difference in optical properties (L versus DL acids).

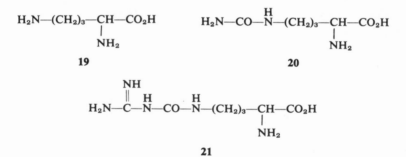

Ito and co-workers (1966) surveyed the distribution of the two novel amino acids in green, brown, and red algae using paper chromatographic techniques. A total of twenty-five algal species was examined. Except for a doubtful spot (visualized by Sakaguchi reagent) for gigartinine (**21**) in one species of brown algae, three species of green and seven species of brown algae failed to contain the two new acids. Of the fifteen species of red algae, however, four species contained gongrine (**18**) and seven species were positive for gigartinine (**21**). Incidentally, none of the algal species that were studied elaborated gongrine without gigartinine. Ito and co-workers (1967) also studied the seasonal variation and the concentration of the two amino acids in marine algae.

Konosu *et al.* (1970), also in Hashimoto's group, have recently reported the isolation of a new betaine, atrinine, from the adductor muscle of a bivalve (Mollusca) *Atrina pectinata japonica* (fan mussel). On the basis of spectral data, including nmr, the structure of atrinine hydrochloride was shown to be trimethyl(2-carboxy-3-hydroxypropyl)ammonium chloride (**22**). This novel betaine is isomeric with the well-known carnitine (**23**).

$$Me_3\overset{+}{N}\text{—}CH_2\text{—}CH\text{—}CH_2OH \quad Cl^- \qquad Me_3\overset{+}{N}\text{—}CH_2\text{—}CH\text{—}CH_2\text{—}CO_2^-$$
$$\qquad\qquad\quad |\qquad\qquad\qquad\qquad\qquad\qquad\qquad\qquad |$$
$$\qquad\qquad CO_2H \qquad\qquad\qquad\qquad\qquad\qquad\qquad OH$$

$$\textbf{22} \qquad\qquad\qquad\qquad\qquad\qquad\qquad\qquad \textbf{23}$$

3. Choline Derivatives

The widely occurring compound choline (**24**), the trimethylammonium cation of ethanolamine, or its simplest ester, acetylcholine (**25**), which is an important intermediate in many biochemical processes, most certainly have no unique or even special place among marine natural products. However, because of the bifunctionality of choline and because the reactive function is a primary and therefore readily esterified alcohol, some of the choline esters that have been isolated from marine sources may well have intrinsic significance or may even prove to be produced exclusively by marine organisms. One of these choline esters in fact is not chemically, but biologically and

$$Me_3\overset{+}{N}\text{—}CH_2\text{—}CH_2OH \quad X^- \qquad Me_3\overset{+}{N}\text{—}CH_2\text{—}CH_2\text{—}OAc \quad X^-$$

$$\textbf{24} \qquad\qquad\qquad\qquad\qquad\qquad\qquad \textbf{25}$$

historically, related to what must be man's oldest economic marine natural product, Tyrian purple (see Section C, this chapter). Briefly (for details *vide infra*), the ancient dye Tyrian purple is produced by marine gastropod mollusks of the family Muricidae. Physiological activity of extracts from the dye-producing glands was discovered early in this century and was later believed to arise from an acetylcholine-like substance. Erspamer (1948) first isolated the substance from *Murex trunculus* in form of several crystalline salts and named it murexine. Its structure (characterized as the picrate), **26**,

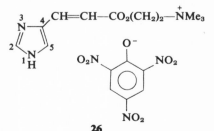

26

was shown to be (Erspamer and Benati, 1953a,b) the choline ester of 3-(4-imidazolyl)acrylic acid by hydrolysis to choline and urocanic acid (27), a degradation product of histidine (28). Structure 26 of murexine was confirmed by synthesis (Pasini *et al.*, 1952).

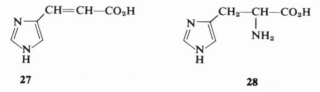

27   28

Whittaker (1960) surveyed a number of marine gastropod mollusks for the occurrence of murexine. He identified murexine as a constituent of all except one member of the Muricidae. *Thais floridana* proved to contain the choline ester of senecioic (β, β-dimethylacrylic) acid (29). Hydrolysis, hydrogenation of the unsaturated acid, and chromatographic comparison with isovaleric acid (30) secured the structure (Keyl *et al.*, 1957), which was subsequently confirmed by synthesis (Whittaker, 1959a). Similarly, from a mollusk that belongs to a family (Buccinidae) that does not secrete a dye—*Buccinum undatum* (common whelk)—Whittaker (1959b, 1960) isolated and identified choline acrylate (31).

$$Me_2C{=}CH{-}CO_2{-}CH_2{-}CH_2{-}\overset{+}{N}Me_3 \quad X^-$$

29

$$Me_2CH{-}CH_2{-}CO_2H$$

30

$$CH_2{=}CH{-}CO_2CH_2{-}CH_2{-}\overset{+}{N}Me_3 \quad X^-$$

31

A choline ester of β-acetoxypalmitic (3-acetoxyhexadecanoic) acid was isolated from a fish and its structure 32 determined by physical measurements, degradation, and synthesis (Boylan and Scheuer, 1967). The fish, *Ostracion lentiginosus* (boxfish), is a member of the trunkfish family (Ostraciontidae). The members of this family are small, colorful, slow-moving reef fishes, not unlike miniature puffer fishes in appearance. This resemblance is not unexpected since trunkfishes and puffer fishes belong to the same order, viz. Plectognathi. Brock (1955) had reported the observation that the boxfish, *Ostracion lentiginosus*, a fairly common inhabitant of reefs in the Hawaiian islands, when placed in an aquarium secretes a substance that kills other fish in its vicinity. Thomson (1963) confirmed this observation and isolated a crude preparation that he believed to be structurally related to the triterpenoid

$$CH_3{-}(CH_2)_{12}{-}\overset{\overset{\textstyle OAc}{\textstyle |}}{C}H{-}CH_2{-}CO_2{-}CH_2{-}CH_2{-}\overset{+}{N}Me_3 \quad X^-$$

32

glycosides of the holothurians (sea cucumbers; see Chapter 1) on the basis of a similarity in foaming and hemolytic properties. Thomson (1964) named his preparation ostracitoxin, while in the meantime Scheuer (1964) had reported the isolation of a pure crystalline toxin that had been designated pahutoxin. (*Pahu* is the Hawaiian word for the boxfish.)

As reported by Boylan and Scheuer (1967), the boxfish were caught by net and were immediately placed in containers of distilled water, where under stress of the changed environment they released copious quantities of a mucous secretion. After this "milking" the boxfish were returned to the sea and the aqueous toxic (capable of killing brackish-water mollies) solution was rapidly extracted with 1-butanol, which after concentration and silicic acid chromatography yielded an amorphous toxin. After passage through an anion exchange column pahutoxin could be crystallized. Hydrolytic degradations under a variety of conditions established structure 32, which was confirmed by synthesis starting with tetradecanol (33) via tetradecanal (34), 3-hydroxyhexadecanoic acid (35), 3-acetoxyhexadeconoic acid (36) to pahutoxin (32), identical with the natural toxin in all properties except optical activity. Boylan and Scheuer (1967) also synthesized the C_{14} and C_{12} homologs of pahutoxin. They compared the hemolytic and lethal properties of the three compounds and found that pahutoxin was the most active of the three, while the C_{12} homolog was the least active and the C_{14} homolog was intermediate in activity.

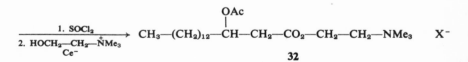

Many marine biologists believe that defensive secretions should be common among defenseless sessile or slow-moving marine forms, and no doubt we may expect many more reports of such substances in the future. The ability of pahutoxin to kill fish may well be related to the substance's ability to act as a detergent. Mann and Povich (1969) carried out surface tension measurements on pahutoxin, but they did not succeed in determining unambiguously the mechanism by which pahutoxin kills fish.

Structurally, the closest relative of pahutoxin that has been isolated from a marine source is the choline ester of 14-methylpentadec-4-enoic acid (**37**), which Nakazawa (1959) isolated from an oyster. The choline ester of **37**, however, is apparently the cation of a complex salt, whose anion is a trisaccharide of the glycopeptide lactyl taurine of structure **38**.

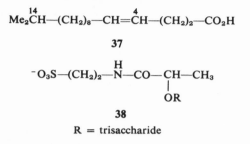

$$\overset{14}{\text{Me}_2\text{CH}}\text{—}(\text{CH}_2)_8\text{—CH}\overset{4}{=}\text{CH}\text{—}(\text{CH}_2)_2\text{—CO}_2\text{H}$$

37

$$^-\text{O}_3\text{S}\text{—}(\text{CH}_2)_2\text{—}\overset{\text{H}}{\text{N}}\text{—CO}\text{—}\underset{\overset{|}{\text{OR}}}{\text{CH}}\text{—CH}_3$$

38

R = trisaccharide

B. Compounds Containing Acyclic Nitrogen

Among the small number of compounds in this group that have so far been isolated from marine organisms and that without a doubt are secondary metabolites are only a few that are related to each other structurally. These are neutral nitrogenous (amide or nitrile) derivatives of similarly substituted acetic acids, perhaps biogenetically derived from a dibromotyrosine. All of these compounds have been isolated from sponges (Porifera). They will be discussed first, followed by a few other unrelated compounds.

2. DIBROMOTYROSIN-DERIVED COMPOUNDS

Burkholder and collaborators in their search for antibiotic substances in marine organisms reported the isolation of four compounds from the sponges *Verongia fistularia* and *V. cauliformis* (Sharma and Burkholder, 1967a; Sharma *et al.*, 1968). Two of the compounds contained bromine and were assigned structures of 2,6-dibromo-4-hydroxy-4-acetamidocyclohexa-2,5-dienone (**39**) and the corresponding dimethylketal (**40**) of compound **39**. Since Sharma and Burkholder (1967a) use methanol as the primary extraction solvent, and since complete absence of acid during the extraction is difficult to judge, it is possible that the dimethylketal **40** is not a natural product,

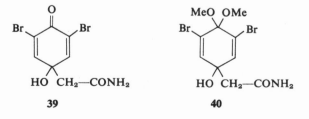

39   **40**

but may be generated during work-up (see below). In their preliminary communication Sharma and Burkholder (1967b) reported spectral data for compound **39** and the outline of a synthesis from 4-hydroxy-3,5-dibromophenylacetamide (**41**). Spectral data for compound **40**, the ketal of **39**, were reported by Sharma *et al.* (1970). In addition, the ketal was readily convertible to its

41

parent ketone **39**, thereby confirming the relationship of the two compounds. The reverse reaction, however, ketalization of the dienone **39**, was reported not to occur under normal conditions of isolation nor by treatment with methyl orthoformate. It would appear then that **40** should be considered a natural product.

Fattorusso and co-workers (1970a, 1972) extracted the sponge *Aplysina aerophoba*, from which they also isolated compound **39** as well as the related compound **42a**, which they named aeroplysinin-1. Its structure was deduced from spectral data of **42a** and its diacetate **42b** and was confirmed by transformation with dilute base to 2-hydroxy-4-methoxy-3,5-dibromophenylacetonitrile **43** and with acid to 3,5-dibromo-2-hydroxy-4-methoxyphenylacetic acid (**44**).

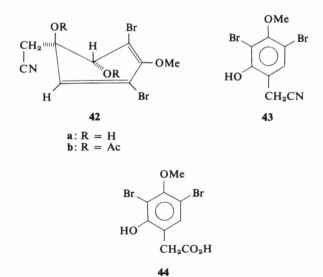

42

a: R = H
b: R = Ac

43

44

Aeroplysinin-1 (**42a**) is dextrorotatory as reported by Fattorusso *et al.* (1970a) who isolated it from the sponge *Aplysina aerophoba*. Remarkably, Fulmor and co-workers (1970) isolated the levorotatory antipode of aeroplysinin-1 from a sponge *Ianthella ardis*. Another (unspecified) *Ianthella* species yielded the dextrorotatory enantiomer. Furthermore, both optical forms exhibited equal antibacterial activity *in vitro*. Fulmor *et al.* (1970) were able to convert aeroplysinin-1 to **39** by treatment with trifluoroacetic acid in acetone. Aqueous acid or alkali treatment of **42a** led to complex mixtures. This inability of **42a** to be readily dehydrated to the aromatic system suggested to Fulmore *et al.* (1970) a trans relationship of the vicinal hydroxyl groups. The Italian workers (Fattorusso *et al.*, 1970a, 1972), as was mentioned, report facile aromatization when **42a** is treated with acid or base. Cosulich and Lovell (1971) carried out a single crystal X-ray determination of the levorotatory isomer and confirmed that the two hydroxy groups in **42a** are indeed *trans* to each other and axial. The X-ray study also confirmed the absolute configuration of aeroplysinin-1 (**42a**), which had been deduced on the basis of circular dichroism measurements by Fulmor *et al.* (1970).

Two additional bromine-containing sponge constituents contain cyclic rather than acyclic nitrogen. However, generically and biologically these two compounds are so closely related to the aeroplysinins that they will be discussed here. Both compounds were isolated by Fattorusso and co-workers (1970b, 1971a) from the same two sponges *Aplysina* (syn. *Verongia*) *aerophoba* and *V. thiona*.

The first, aerothionin, is the major constituent of the two sponges and was shown to have the bisisoxazole structure **45**. Structure **45** was deduced on the

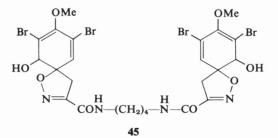

45

basis of spectral properties and was confirmed by the following transformations. Treatment of aerothionin (**45**) with dilute alkali yielded compound **46a**, in which the dihydrobenzene rings are aromatized and the isoxazole rings have been opened to form oximes. This phenol dimethyl ether (**46a**) was transformed to hexamethyl ether **46b**, which upon treatment with 25% aqueous ethanolic potassium hydroxide furnished the phenylpyruvic acid derivative **47**, which was synthesized. Furthermore, reaction of the acid chloride of **47** with 1,4-diaminobutane resulted in **46b**.

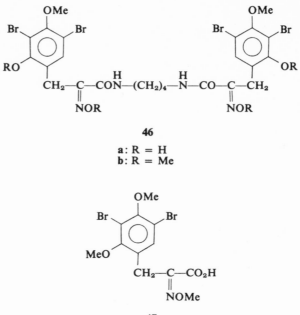

46

a: R = H
b: R = Me

47

From the mother liquors of aerothionin (**45**) Fattorusso *et al.* (1971a) isolated a homolog, homoaerothionin (**48**), the structure of which was established in full analogy with that of aerothionin. A full account of this work is available (Moody *et al.*, 1972).

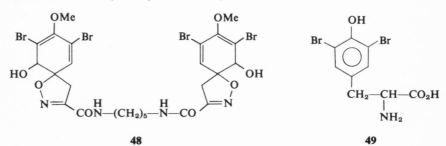

48   **49**

Fattorusso and co-workers suggest that the aerothionins (as well as aeroplysinin-1) are biosynthesized from a derivative of 3,5-dibromotyrosine (**49**), which condenses with an appropriate α,ω-diaminoalkane derived from the corresponding amino acids ornithine (**19**) or lysine (**50**).

$$\overset{\displaystyle NH_2}{\underset{\displaystyle}{|}}$$
$$H_2N{-}(CH_2)_4{-}CH{-}CO_2H$$

50

2. CAULERPICIN

A substance containing neutral acyclic nitrogen in a long-chain saturated hydrocarbon was first isolated by Doty and Santos (1966) from the green alga *Caulerpa racemosa* and was named caulerpicin. The algal genus *Caulerpa* is known in the Philippines for its peppery taste and some species are edible and cultivated. On the basis of spectral data Santos and Doty (1968) proposed structure **51** for caulerpicin. Mass spectral data indicate that caulerpicin may be a mixture of homologs. Full confirmation of structure **51** has not yet been reported.

$$CH_3—(CH_2)_{13}—\overset{\displaystyle CH_2OH}{\underset{\displaystyle}{CH}}—\underset{\displaystyle H}{N}—CO—(CH_2)_n—CH_3$$

51

$n = 23, 24, 25$

3. NEREISTOXIN

It had been known in Japan that some insects die when they come in contact with a common fish bait, the marine annelid *Lumbriconeris heteropoda* (Hashimoto and Okaichi, 1960). The toxic principle could be isolated as a hydrogen oxalate salt and was named nereistoxin. Its full structural elucidation (Okaichi and Hashimoto, 1962a) leading to **52**, 4-dimethylamino-1,2-dithiolane, became possible when the bisbenzoyl derivative (**54**) of the initial sodium borohydride reduction product (**53**), upon treatment with Raney nickel yielded dimethyl isopropylamine (**55**). The physiological activity of nereistoxin as an insecticide (Okaichi and Hashimoto, 1962b) made the synthesis of nereistoxin (**52**) a desirable objective. It proved to be a difficult undertaking considering the simplicity and symmetry of the structure, but was accomplished by Hagiwara *et al.* (1965). The difficulties encountered are a direct consequence of the presence in the molecule of two sensitive functional groups that are separated by only three carbon atoms.

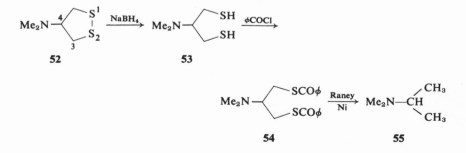

For the synthesis of nereistoxin (52) Hagiwara *et al.* (1965) reacted 1,3-dichloro-2-propanol (56) with sodium benzylmercaptide (57) to yield 1,3-bis(benzylthio)-2-propanol (58). Treatment of 58 with thionyl chloride led to a mixture of the normal (and hoped-for) substitution product 59a and an (unexpected) rearranged isomer 60a. The structures of 59a and 60a were proven by spectral data and by further chemical transformations. The unseparated mixture of 59a and 60a was heated under pressure with dimethylamine in benzene and furnished a mixture of the corresponding dimethylamino derivatives 59b and 60b. The asymmetrical (and undesired) compound 60b was the major product, which could be isolated and crystallized as the hydrogen oxalate salt. Identity of the minor amine 59b was secured by chromatographic comparison with a degradation product of nereistoxin. Birch reduction of the mixed amines 59b and 60b led to a mixture of amine dithiols (61 and 62), which was readily separable since the symmetrical

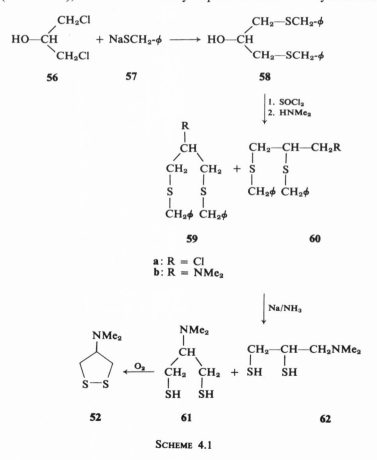

SCHEME 4.1

compound **61** could be extracted into ether from an alkaline solution. Moreover, the desired dihydronereistoxin (**61**) was being oxidized to nereistoxin (**52**) spontaneously during work-up. The yield of nereistoxin (**52**) from the mixed amines **59b** and **60b** was 6%. Scheme 4.1 summarizes the synthesis.

C. Compounds Containing Nitrogen in a Cycle

During the discussion of the choline ester murexine (see Section A,3) brief reference was made to the ancient dye Tyrian purple, derived from a few molluscan species of the genus *Murex*. Tyrian purple proved to be a derivative of the the the nitrogen heterocycle indole (**63**), which in turn consists of the nitrogen heterocycle pyrrole (**64**) fused to benzene. The marine natural products that involve the indole or pyrrole system will be presented first, followed by a few examples of compounds with six-membered nitrogen heterocycles, and finally by some complex systems.

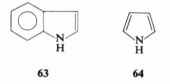

63   **64**

1. DERIVATIVES OF INDOLE AND PYRROLE

In a recent account of the history of Tyrian purple Robinson (1971) remarked that Tyrian purple probably was one of the most expensive commodities of the ancient world. According to this estimate a silk scarf dyed in Tyrian purple would have cost the equivalent of $900 in A.D. 300. This high cost was caused by the fact that one animal collected by the Phoenicians yielded only two drops of dye-producing secretion and that it took a minimum of 60,000 mollusks to produce one pound of Tyrian purple.

The *Murex* secretion is described as a yellow-white creamy fluid with a strong odor of garlic (Robinson, 1971). When this fluid is placed on wool or linen and exposed to strong light it will undergo a series of color changes, ending up in a deep purple-red. After the fabric is washed in soap and water, the dye becomes a bright crimson, which is color fast. This dye production from mollusks is believed to go back at least to 1600 B.C., when the process was discovered by the Cretans. By the year 1000 B.C. wool and silk dying had become a thriving business in the ancient city of Tyre in the Eastern Mediterranean.

Tyrian purple thus is no doubt the oldest commercial marine natural product. It also is the earliest marine metabolite whose structure was correctly deduced and proven. Friedländer (1907) in a preliminary communication reported on the isolation of the secretion of *Murex brandaris* and of the production of the dye after exposure to light. He also showed that Tyrian purple was different from the known dyes indigo and thioindigo, but that it was related to these compounds. During the summer of 1908 Friedländer worked up 12,000 snails (*Murex brandaris*) and isolated 1.4 g of analytically pure dye (Friedländer, 1909). Of the twenty-two possible dibromoindigotins Friedländer (1909) correctly excluded the eighteen asymmetrical compounds on the basis of his observation that the colorless precursor of the dye was readily soluble, from which he shrewdly concluded that the purple dye was likely to be a symmetrical dimer of the colorless precursor. He had thus reduced the structural possibilities from twenty-two to four! He had insufficient material for degradation of the dye—a statement that is readily apparent when one considers that his elemental analyses for carbon, hydrogen, nitrogen, and bromine (duplicate only of bromine) had burned up no less than 0.5 g of his 1.4-g supply. Two of the four symmetrical compounds had been described at the time—the 5,5'-dibromo isomer had been prepared by Baeyer (1879) and 6,6'-dibromoindigotin (**65**) had been synthesized by Sachs and co-workers (Sachs and Kempf, 1903; Sachs and Sichel, 1904).

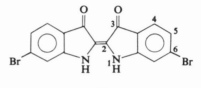

65

On the basis of the published properties of the two compounds (solubilities and color appearance in solution) Friedländer concluded that Sachs' 6,6'-isomer was the better choice. He synthesized this compound by a new route from 2-amino-4-bromobenzoic acid (**66**). Synthetic and natural dyes proved to be identical not only by the normal criteria of the day, solubility and color tests, but in their visible absorption spectra.

It had been recognized from the earliest times that the purple dye 6,6'-dibromoindigotin (**65**) was not the secretion of the mollusk, but that a colorless precursor was the true metabolite. Yet aside from an investigation by Bouchilloux and Roche (1954a,b; 1955) that achieved an isolation of the colorless precursor and a characterization of its functional groups no structural work seems to have been carried out until recently (Baker and Sutherland, 1968). Baker and Sutherland (1968) used the mollusk *Diacathais orbita* and from an extraction of the dye (hypobranchial) glands were able to separate the silver salt of 6-bromo-2-methylmercaptoindoxyl-3-sulfate (**67a**) as a crystalline entity. Structure **67** was deduced from an nmr spectrum of the potassium salt (**67b**) and by degradation with Raney nickel to indoxyl sodium sulfate (**68**), identical with a synthetic sample. Baker and Sutherland

(1968) were also able to extract from the molluscan glands an enzyme concentrate, which was capable of converting the sodium salt of the precursor (**67c**) to the purple dye on a paper chromatogram, as shown by identical R_F values.

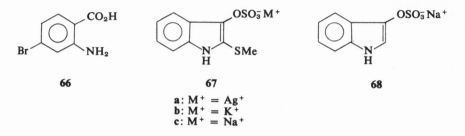

66   67   68

a: $M^+ = Ag^+$
b: $M^+ = K^+$
c: $M^+ = Na^+$

When a moist ether extract of freshly excised glands of *D. orbita* was treated with diazomethane at 0°, two additional crystalline compounds were isolated by Baker and Sutherland (1968). On the basis of spectral data structures **69** and **70** are proposed. If the ether extract is allowed to stand, the pale yellow-green tyriverdin of composition $C_{36}H_{30}Br_4N_4O_5S_4$ is produced. Baker and Sutherland (1968) suggest a quinhydrone type structure for tyriverdin on the basis of spectral data and the results of acetylation with acetic anhydride and perchloric acid at 0°C.

69   70

Structure **67** for the precursor of Tyrian purple thus not only fits all the known chemical facts, but even explains the strong garlic odor that is mentioned in the old literature.

The only other simple derivative of indole (**63**) that has been reported to date from a marine source is a substance with antibiotic activity that Stempien (1966) isolated from several species of sponges of the genus *Agelas* and to which he, largely on the basis of color tests, assigned either structure **71** (4,6-dihydroxyindole) or **72** (6,7-dihydroxyindole).

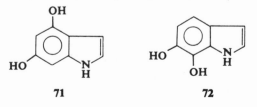

71   72

Perhaps the most intriguing of the derivatives of pyrrole (**64**) is a phenyl-substituted pyrrole with no fewer than five bromine atoms that Burkholder *et al.* (1966) isolated from a marine bacterium *Pseudomonas bromoutilis.* Lovell (1966) determined its structure by single crystal X-ray techniques to be 2-(2-hydroxy-3,5-dibromophenyl)-3,4,5-tribromopyrrole (**73**). The compound, which exhibited *in vitro* antibiotic activity, was synthesized by

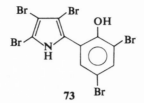

73

Hanessian and Kaltenbronn (1966) by condensing 3,5-dibromo-2-methoxy-acetophenone (**74**) with 1-nitro-2-dimethylaminoethylene (**75**) in the presence of base to yield an intermediate **76**, which reductively cyclized to form 2-(3,5-dibromo-2-methoxyphenyl)pyrrole (**77**). Compound **77** on bromination and demethylation with boron trichloride in carbon tetrachloride furnished the antibiotic **73**. This sequence of reactions is summarized in Scheme 4.2.

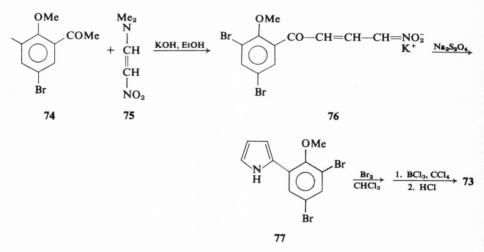

SCHEME 4.2

Three simple and related bromopyrroles have recently been isolated by the Italian group (Forenza *et al.*, 1971) from the sponge *Agelas oroides* in yields ranging from a trace to 0.2% based on dry weight of the animals. The compounds were shown by spectral data and interconversions to be 4,5-dibromopyrrole-2-carboxylic acid (**78a**) and the corresponding amide (**78b**) and nitrile (**78c**).

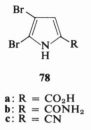

78

a: R = CO$_2$H
b: R = CONH$_2$
c: R = CN

Several related amino acids, derivatives of tetrahydropyrrole (pyrrolidine), were isolated from the red alga *Digenea simplex* as the active principles of an algal species known and used for its anthelmintic properties. The principal and most active constituent was characterized by Murakami and co-workers (1953) and was first called digenic acid. Murakami *et al.* (1954) subsequently proposed the trivial name α-kainic acid, which has been continued. In a series of classic chemical degradations and syntheses of degradation products several groups of Japanese workers (Honjo *et al.*, 1955; Murakami *et al.*, 1955a; Nawa *et al.*, 1955a,b; Sugawa *et al.*, 1955a,b; Ueno *et al.*, 1955) delineated the structure of α-kainic acid as 3-carboxymethyl-4-isopropenyl-pyrrolidine-2-carboxylic acid (**79**). Morimoto (1955) deduced the relative stereochemistry as shown in structure **79**. Among the early and important degradative reactions were a soda-lime distillation that led to the isolation of 3-isopropylpyrrole (**80**), ozonization yielding formaldehyde, and chromic acid oxidation that furnished 3-carboxymethyl-4-isopropylmaleimide (**81**).

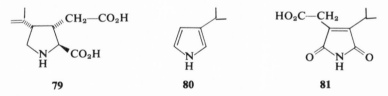

79   **80**   **81**

From the mother liquors of α-kainic acid (**79**) Murakami and collaborators (1955b) isolated a stereoisomer of α-kainic acid, which they designated α-allokainic acid. Murakami *et al.* (1955c) recognized that the new isomer differs from α-kainic acid only by the configuration of the isopropenyl group and assigned structure **82** to α-allokainic acid.

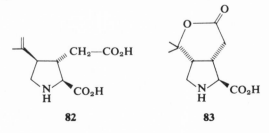

82   **83**

A third isolate, designated kainic lactone and assigned structure **83**, is considered an artifact.

A structurally related anthelmintic, domoic acid, was isolated by Daigo (1959) from another red alga, *Chondria armata*. Its structure, differing from the kainic acids by an octadienoic acid instead of an isopropenyl side chain, was determined to be L_s-arabo-2-carboxy-4-(1-methyl-5-carboxy-*trans,trans,s-trans*-1,3-hexadienyl)-3-pyrrolidine acetic acid (**84**) by a combination of classical degradations and spectral techniques (Takemoto *et al.*, 1966).

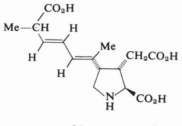

84

Among the two major groups of pigments based on pyrrole the macrocyclic tetrapyrroles (e.g., the heme and most chlorophyll pigments) occur universally and fall therefore outside the scope of this book. The story of the linear tetrapyrrole pigments (the bile pigments or bilirubinoids) is less clear. According to With's (1968) authoritative monograph, in humans and a few other vertebrates that have been studied thoroughly, the bile pigments are considered biological degradation products. In invertebrates, however, the distribution of bile pigments is irregular and it is not known whether they have a physiological function. In plants the bile pigments were believed to be restricted to red and blue-green algae and legumes, but they have now been recognized as prosthetic groups of the chromoprotein phytochrome and are therefore widely distributed.

Whether marine invertebrates elaborate any unique linear tetrapyrrole pigments is uncertain at the present time. Almost all of the early work was based on color reactions, particularly the Gmelin test—a treatment of the pigment material with fuming nitric acid and the observation of a series of successive color changes. Only rarely was crystalline pigment isolated and structural studies have appeared only recently.

For many years, for example, the pigment that occurs in the calcareous skeleton of the blue coral *Helipora coerulea* was believed to be a unique bile pigment and was referred to as helioporobilin. Rüdiger and co-workers (1968) have recently reisolated this pigment and have shown by thin-layer chromatography and mass spectrometry that the major constituent of the pigment mixture was the known biliverdin-IX of structure **85**.

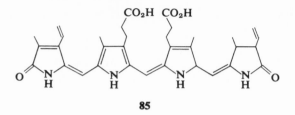

85

The indefinite status of bile pigments from invertebrates may be illustrated by another example. Sea hares (gastropod mollusks) have long had a reputation of being toxic and some of them have also been known to discharge a purple secretion, generally believed to have a defensive character. Flury (1915) showed that a milky secretion of the sea hare *Aplysia depilans* was indeed toxic to a number of marine invertebrates, but that the purple secretion of *A. limacina* was nontoxic. Attempts of chemical definition of the pigment, then called aplysiopurpurin, began in 1925. Only recently, Rüdiger (1967a,b) reisolated aplysioviolin from *A. limacina* and by chemical degradation and spectral data proposed structure **86a**, presumably a unique but minor variant of the bile pigments.

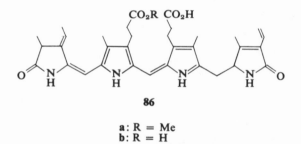

86

a: R = Me
b: R = H

The parent compound of aplysioviolin (**86a**) is the corresponding free acid (**77b**); it bears the trivial name phycoerythrobilin and is an algal chromophore (Chapman *et al.*, 1967). The close structural relationship of the two pigments led Chapman and Fox (1969) to check the hypothesis whether aplysioviolin (**86a**) in *Aplysia californica* is derived from phycoerythrobilin contained in a diet of red algae. This proved to be the case. Chapman and Fox's (1969) experiments further indicated that the purple secretion of sea hares may be a metabolic end product without defensive function.

2. Six-Membered Nitrogen Heterocycles

A suggestion that two neurotoxic nicotine alkaloids may be constituents of members of the invertebrate phylum Rhynchocoela was first made by Bacq (1937) and was based on the observation that saline extracts of whole nemertine worms when injected into crabs produced convulsions, paralysis, and death. Bacq (1936) coined the trivial names amphiporine and nemertine for these toxins.

Solution of this problem was not forthcoming until 1969 in a preliminary announcement by Kem *et al.*, followed by the full paper in 1971. Kem *et al.* (1971) collected and processed about ten thousand (3–4 kg) specimens of the hoplonemertine *Paranemertes peregrina*. The toxin was purified by solvent partition, chromatography, and preparation of the crystalline picrate. Spectral data and chemical reduction suggested the structure of the toxin to be 2-(3-pyridyl)-3,4,5,6-tetrahydropyridine or anabaseine (**87**). Whereas the dihydro derivative of **87**, anabasine (**88**), is a well-known naturally occurring plant alkaloid, the nemertine toxin anabaseine (**87**) was only known in the literature as a synthetic product (Späth and Mamoli, 1936). Kem *et al.* prepared synthetic anabaseine (**87**) and confirmed its structure by direct comparison.

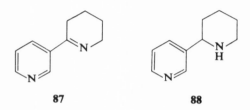

87   **88**

Kem (1971) also studied the distribution of anabaseine in nemertine tissue extracts as well as in the phylum. The nemertines are carnivores that are capable of capturing and ingesting prey several times their own size, presumably by release of a venom from their proboscis. It was therefore interesting to discover (Kem, 1971) that in *Paranemertes* the anterior proboscis contained only 27%, while the body proper (largely in the integument) accounted for 69% of the total anabaseine. Kem (1971), however, estimates that the anterior proboscis contains about seventy times the anabaseine necessary to paralyze an annelid its own size. Among the four orders of nemertines, anabaseine was found in only one, the Hoplonemertinea. In this order anabaseine was found in three of five genera that were surveyed. Altogether, however, only thirteen species, constituting only about 2% of the described species were investigated.

The only other simple relative of plant alkaloids so far reported from a marine source was recently discovered by Fattorusso *et al.* (1971b) in aqueous residues of the sponge extracts (*Aplysina*, syn. *Verongia aerophoba*) that had yielded several bromotyrosine-derived constituents (*vide supra*). The substance was present in high yield (2.5% of dry animal) and was shown by spectral data, degradation, and synthesis to be 3,4-dihydroxyquinoline-2-carboxylic acid (**89**). Like anabaseine (**87**), this quinoline derivative also was not previously known as a natural product.

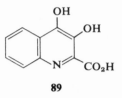

89

The third compound in this group, a pyrazine derivative, is more complex than the two pyridine compounds. It was isolated from the green alga *Caulerpa racemosa* (Santos and Doty, 1968) in an investigation that was aimed at the toxic and flavor constituents of *Caulerpa*, of which some species are eaten as a salad in the Philippines. The heterocycle, named caulerpin, is a red substance and was crystallized from ether extracts of the algae (Santos, 1970). Spectral data and a few degradative reactions pointed to a structure of dimethyl 6,13-dihydrodibenzo[*b,i*]phenazine-5,12-dicarboxylate (**90**) for caulerpin, which has not yet been further confirmed by synthesis or single-crystal X-ray data.

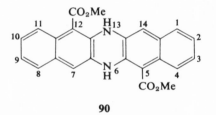

90

In addition to these six-membered nitrogen heterocycles there are literature reports of the isolation or at least identification of simple purine and pteridine derivatives from marine organisms. As has been pointed out earlier, these compounds are not within the scope of this book, unless they constitute a distinct variant of the run-of-the-mill purine or pteridine (see, e.g., saxitoxin in the following section).

D. Complex Polycyclic Compounds

1. THE PHAKELLINS AND OROIDIN

From the sponge *Phakellia flabellata* Sharma and Burkholder (1971) isolated two complex nitrogen heterocycles, dibromophakellin (**91a**) and

4-bromophakellin **(91b)**. The structures were deduced on spectral evidence and confirmed by single-crystal X-ray analysis of a monoacetate.

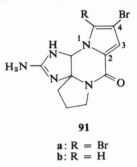

91

a: R = Br
b: R = H

A compound, which may well be a biological precursor of the phakellins was recently isolated by Forenza *et al.* (1971) from another sponge *Agelas oroides*. This sponge yielded small quantities of three simple bromopyrrole-carboxylic acid derivatives (see Section C,1). A more complex derivative was present in far greater concentration (2.3%). It was named oroidin and proved to be an imidazole amide of the bromopyrrole carboxylic acid of structure **92**. Spectral data and hydrolytic degradations as well as synthesis of the hydro-lytic fragments secured the structure.

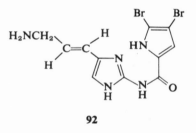

92

2. SAXITOXIN

Paralytic shellfish poisoning or mussel poisoning has been known from many parts of the world for a long time, but only during the past thirty or so years have the source, the epidemiology, the physiological action, and the chemical nature of the poison that causes the outbreaks become at least well understood if not fully elucidated. Schantz, one of the active workers in the field, has ably reviewed the area of shellfish poisons from time to time (Schantz 1960, 1969, 1971). These articles should be consulted for additional details.

The geographical areas that are most frequently associated with paralytic shellfish poisoning are the North Atlantic coasts of Europe and America, the North Pacific coast of America from California to Alaska, and the coastal areas of Japan and South Africa. Outbreaks occur suddenly and irregularly for no apparent reason and generally last for a few weeks. Early human symptoms that may appear within thirty minutes after eating toxic shellfish consist of numbness of lips and fingertips; this is followed by paralysis and death, generally within 1–12 hours depending on the dose. It is estimated that the lethal dose in man is about 0.3 mg of toxin.

It was only in 1937 that a group of researchers in California (Sommer and Meyer, 1937; Sommer *et al.*, 1937) connected the presence of a dinoflagellate *Gonyaulax catenella* in California waters to the outbreak of mussel poisoning. It was established that a *G. catenella* count of 100 to 200 cells per milliliter of water rendered the mussels too toxic for human consumption. However, a visible "red tide" is achieved only when the organism reaches or surpasses a count of 20,000 cells per milliliter. The threshold count of about 200 cells per milliliter is detectable only by microscopic examination. Obviously then, paralytic shellfish poisoning continues to be a public health threat and the popularly held danger signal of the "red tide" misses being an alarm beacon by two orders of magnitude!

Although there existed considerable evidence that the toxin that was elaborated by the dinoflagellate *G. catenella* was the same as the toxin that is accumulated in the dark gland or hepatopancreas of the mussel *Mytilus californianus*, identity was only conclusively demonstrated in 1966. Schantz and co-workers (1966) isolated the toxin from an axenic culture of *G. catenella* and from the dark gland of *M. californianus* and proved their identity. Schantz *et al.* (1966) further showed that the poison is also identical with saxitoxin, the toxin accumulated and stored in the siphon of the Alaska butter clam *Saxidomus giganteus*, and which had been characterized and named by Schuett and Rapoport (1962). In contrast to the mussel, which becomes safe for human consumption after a few weeks, the Alaska butter clam retains the toxin in its siphon for many months. Furthermore no direct link has been established between the toxicity of *S. giganteus* and another toxic organism. Schantz and Magnusson (1964) did locate *Gonyaulax* in Alaskan waters, but only rarely and in small numbers.

No fully conclusive proof of the identity of saxitoxin with the paralytic shellfish poison produced by the Bay of Fundy scallop *Pecten grandis* has yet been published (Schantz, 1960), but it has been shown that toxicity of Bay of Fundy shellfish is associated with plankton blooms, particularly of *G. tamarensis* (Needler, 1949).

Only a few papers have been published on the chemistry of saxitoxin, which with an intravenous lethal dose of 3–4 μg/kg (Murtha, 1960) is one of the most toxic nonprotein substances known. Isolation and purification were described by Schantz *et al.* (1957) and Mold *et al.* (1957), who established the molecular formula of saxitoxin dihydrochloride as $C_{10}H_{17}N_7O_4 \cdot 2\,HCl$ and demonstrated its homogeneity although neither saxitoxin nor its salt could be crystallized. Among the properties that were reported for saxitoxin were its catalytic reduction to a nontoxic dihydro derivative (Mold *et al.*, 1957); its two basic functions of pK_a 8.1 and 11.5 (Schantz *et al.*, 1961); and its drastic oxidative degradation (Schantz *et al.*, 1961) with periodic acid or permanganate that yielded ammonia, carbon dioxide, urea, and guanidinopropionic acid (**93**).

$$\underset{\text{H}}{\underset{|}{\text{H}_2\text{N}-\overset{\overset{\text{NH}}{\|}}{\text{C}}-\text{N}-\text{CH}_2-\text{CH}_2-\text{CO}_2\text{H}}}$$

93

In a significant degradation reported by Schuett and Rapoport (1962) saxitoxin was treated with phosphorus and hydriodic acid in acetic acid leading in high yield to a crystalline weakly basic substance that was shown to be 8-methyl-2-oxo-2,4,5,6-tetrahydropyrrolo[1,2c]pyrimidine (**94**).

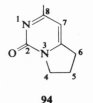

94

Schuett and Rapaport (1962) synthesized compound **94**, which contained eight of the ten carbon atoms of saxitoxin and which constituted a new heterocyclic system, by the following route. Pyrrole-2-aldehyde (**95**) was condensed with nitroethane (**96**). The resulting product **97** was successively reduced to **98** and then **99**. Pyrrolidine **99** was heated with diethyl carbonate and yielded 8-methyl-2-oxo-1,2,4,5,6,6a,7,8-octahydropyrrole[1,2c]pyrimidine (**100**), as outlined in Scheme 4.3.

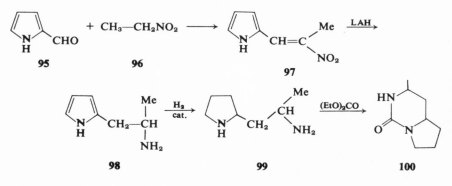

SCHEME 4.3

Compound **100** could be oxidized with permanganate to the saxitoxin degradation product **94**, which in turn could be hydrogenated to its tetra-hydro analog **100**. Both synthetic compounds **100** and **94** were identical with the corresponding degradation products of saxitoxin.

Russell (1967) in his review published **101** as the structure of saxitoxin and credits it to Rapoport *et al.* (1964). This structure has since been widely quoted (e.g., Baslow, 1969) but it should be noted that the composition of **101** ($C_{10}H_{15}N_7O_3$) differs from that of saxitoxin by the elements of water and that Russell's (1967) reference to Rapoport *et al.* (1964) must be to an oral presentation since the published abstract contains no structural formula. Furthermore, recent work by Rapoport's group (Wong *et al.*, 1971a,b) shows this structure to be incorrect (*vide infra*).

In a recent publication from Rapoport's group (Wong *et al.*, 1971a) details of further degradative studies on saxitoxin have been revealed. Although the earlier degradation product **94** was the result of reductive reaction, the new compound **102** was isolated as a crystalline salt following

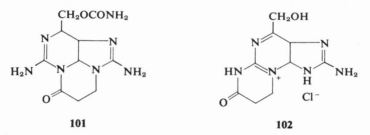

101   **102**

mild oxidation with alkaline hydrogen peroxide. This new degradation product lacks only one carbon and one nitrogen atom of the intact saxitoxin molecule. By heating **102** in alkali this compound could be further degraded to the salt of the purine derivative **103**. Phosphorus and hydrogen iodide reduction of **103** yielded the deoxy derivative **104**. The structures of the compounds **103** and **104** were deduced on spectral grounds and confirmed by synthesis.

The principal difference between purine derivative **103** and the saxitoxin degradation product **102** is the three-carbon unit derived from propionic

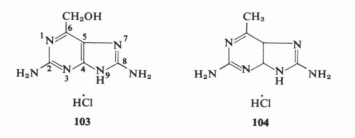

103   **104**

acid. The correct site for attachment of this unit was discovered when mild phosphorus and hydriodic acid treatment of **102** led to a substance **105** having an opened lactam ring. Structure **105** was also confirmed by synthesis.

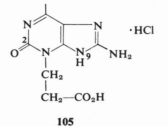

105

The remaining question, whether the second terminus of the propionyl side chain was linked to the nitrogen atom attached to C-2 or to N-9 of the purine system, was decided in favor of the C-2 nitrogen on the basis of yet a further degradation. When compound **102** was successively treated with diazomethane, alkali, and phosphorus-hydriodic acid, compound **106** was isolated and its correct structure was established by synthesis.

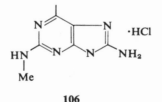

106

In a further communication from Professor Rapoport's laboratory (Wong *et al.*, 1971b) determination of the complete structure of saxitoxin has been reported to be **107**. A remarkable feature of this unique molecule is a carbon (shown by asterisk) that is linked to two nitrogen and two oxygen atoms. This achievement ranks as one of the significant milestones in the chemistry of marine natural products and brings to successful conclusion a difficult structural investigation that had its beginnings some thirty years earlier.

Occurrence of saxitoxin or substances closely related to it has been reported from two other organisms. Jackim and Gentile (1968) isolated from

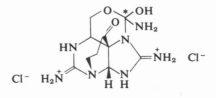

107

a culture of the blue-green alga *Aphanizomenon flos-aquae* a toxin that was nearly identical with saxitoxin in a number of physical and physiological parameters.

The second isolation resulted from an epidemiological investigation of toxic crabs in the Ryukyu and Amami islands by Hashimoto and co-workers (1967a). A number of species were screened and from toxic specimens of the crab *Zosimus aeneus* a toxin was isolated that is almost certainly identical with saxitoxin (Konosu *et al.*, 1968; Noguchi *et al.*, 1969).

While there is no way of detecting the presence of saxitoxin in infested shellfish except by bioassay in mice, there have been, from time to time, reports of massive fish kills that were presumably caused by dinoflagellates. One such occurrence was reported by Hashimoto and co-workers (1968), who isolated the toxin produced by the planktonic organism *Peridinium polonicum* that had been responsible for fish mortality in Lake Sagami near Tokyo in 1962. Hashimoto *et al.* (1968) were able to prepare a crystalline though unstable reineckate of the toxic base which they called glenodinine. It appeared from spectral data and spot tests that glenodinine is an alkaloidal substance possessing a substituted piperidine moiety and a sulfhydryl group.

3. TETRODOTOXIN

a. Introduction

Interest in another potent marine toxin, tetrodotoxin, which along with saxitoxin is a complex nitrogen heterocycle, was originally generated because tetrodotoxin, too, presented a public health hazard, having a lethal dose of the same order of magnitude as saxitoxin. Tetrodotoxin also proved to be a relatively small, basic molecule of limited solubility in organic solvents. On the other hand, these superficial similarities are matched by considerable differences. Poisoning by tetrodotoxin has long been traced to members of a single family of fishes, the Tetraodontidae, and only the viscera, largely the ovaries and the liver, harbor the toxin. Geographically, puffer fish poisoning is known only from Japan, where puffer fish (*fugu*) is a culinary delicacy and where elaborate precautions are taken to prevent human intoxication. Halstead (1967) has described the *fugu* cult in great detail. Briefly, only trained and licensed personnel may handle the fish in the market and remove the viscera. In spite of these precautions 500 intoxications were reported in Japan in 1958–1959, of which 294 proved to be fatal (Tsuda, 1966).

Professor Tsuda, the foremost pioneer in the structural elucidation of tetrodotoxin, writes (Tsuda, 1966) that 2000-year-old Chinese medical books report of puffer fish poisonings and that *fugu* bones have been found in Japanese burial sites that are some fifteen hundred years old. Yet the earliest reports in the Japanese literature on *fugu* intoxication date back only about four hundred years.

Tsuda (1966) also documents the seasonal variation of *fugu* toxicity, which may well be related to the sexual cycle of the puffers. Toxicity is highest during the winter months, which coincides with the prespawning period as well as the time when the fish tastes best. But even during the time of greatest toxicity only about half of the tested specimens were toxic. Of some thirty-seven species of puffers that are recognized around Japan only about six are considered food fishes and among these torafugu, *Fugu rubripes rubripes*, is held in highest esteem. Because of ever more effective control measures intoxications and fatalities have been steadily decreasing in Japan.

Although an impure toxic fraction, then called tetrodotoxin, was first obtained by Tawara (1909), pure crystalline toxin, then named spheroidine, was isolated only in 1948 by Yokoo. Tsuda and Kawamura (1952) crystallized the toxin independently and resumed the earlier name tetrodotoxin. This was the beginning of Tsuda's extensive researches on the structure of tetrodotoxin. Other groups that carried out independent structural investigations were those of Professor Hirata at Nagoya, Professor Woodward at Harvard, and Professor Mosher at Stanford. All four groups concluded the structural research successfully and about at the same time. It was a nice and fitting coincidence that the full details of the tetrodotoxin structure were first disclosed in the spring of 1964 at the Third International Symposium on the Chemistry of Natural Products held in Kyoto, Japan. All four groups have published details of their researches (Tsuda *et al.*, 1964; Tsuda, 1966; Goto *et al.*, 1965; Woodward, 1964; Mosher *et al.*, 1964).

b. Structure

Tetrodotoxin is a colorless weakly basic optically active, crystalline substance that is virtually insoluble in all but acidic media. Its exact elemental composition, $C_{11}H_{17}N_3O_8$, was in doubt for a long time because the toxin retains moisture and solvents easily and effectively. Its composition shares with saxitoxin the interesting property that the number of oxygen and nitrogen atoms equals or exceeds the number of carbon atoms. The composition of the two marine toxins differ from each other, however, in that saxitoxin is heavily nitrogenous while tetrodotoxin is highly oxygenated.

Woodward (1964) demonstrated that the three nitrogen atoms of tetrodotoxin are present in the molecule as a guanidine moiety by isolating guanidine picrate in yields exceeding 30% following vigorous oxidation of the toxin with aqueous sodium permanganate at 75°.

A wide variety of drastic degradations (Woodward, 1964; Tsuda *et al.*, 1962a,b,c)—warm aqueous sodium hydroxide, pyridine-acetic anhydride followed by vacuum pyrolysis, phosphorus and hydrogen iodide followed by potassium ferricyanide, and concentrated sulfuric acid—all yielded closely related quinazoline derivatives of structure **108**, where the nature of R and

R′ depends on the mode of degradation. These key quinazoline compounds confirmed the three nitrogen atoms as a guanidine moiety and indicated strongly that six of the eleven carbon atoms of tetrodotoxin are part of the carbocyclic ring.

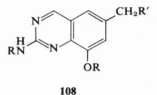

108

The clearly demonstrated guanidine function gave rise to an interesting structural puzzle. The observed weak basicity of tetrodotoxin (pK_a 8.5) is too low to be associated with a normally strongly basic guanidine system. In accordance with the basic nature of the toxin, many attempts to prepare crystalline salts of tetrodotoxin were made but all failed. However, an unusual combination of 0.2 N hydrogen chloride in methanol–acetone did yield a crystalline derivative to which a structure of *O*-methyl-*O*′,*O*″-isopropylidenetetrodotoxin hydrochloride monohydrate (**109**) was assigned on the basis of a single-crystal X-ray analysis (Woodward, 1964). The molecular formula of **109** is $C_{15}H_{23}N_3O_8 \cdot HCl$. If one subtracts from this formula hydrogen chloride (HCl), methanol (CH_4O), and acetone (C_3H_6O), or an aggregate of $C_4H_{11}ClO_2$, and if one then adds the two molecules of water (H_4O_2), which are eliminated in the formation of methyl ether and acetonide functions, one arrives at a net change of C_4H_7Cl. If one continues this atomic arithmetic and deducts this composite change of C_4H_7Cl from $C_{15}H_{23}N_3O_8 \cdot HCl$, one arrives at $C_{11}H_{17}N_3O_8$, precisely the molecular formula of tetrodotoxin. Although this kind of formal atomic arithmetic is not necessarily simply related to real chemical transformations, it is tempting to speculate that the derivative **109** bears a close structural relationship to tetrodotoxin. A comparison of the nmr spectra of the two compounds substantiates the above speculations.

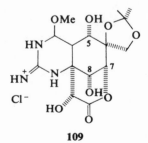

109

Tetrodotoxin derivative **109** differs, however, from the toxin itself in one important structural feature. Compound **109** is a lactone possessing the requisite ir band at 1751 cm⁻¹, but tetrodotoxin (**111**) is devoid of lactonic absorption. On the other hand, the infrared bands assigned to guanidine (1658, 1605 cm⁻¹) remain unchanged as one proceeds from tetrodotoxin (**111**) to derivative **109**. This indicated that the new crystalline hydrochloride (**109**) cannot be a guanidinium salt of tetrodotoxin (**111**). As has been pointed out, tetrodotoxin (**111**) is too weak a base for its basicity (pK_a 8.5) to be associated with the guanidine moiety of the molecule. Derivative **109** is an equally weak base, having a pK_a in water of 8.3. The pK_a of **109**, however, is 9.2 when the measurement is made in dioxane water. This is a significant finding since enhanced pK_a values are characteristic of hydroxyl group dissociation when one proceeds from solvents of high to solvents of low dielectric constant. From this series of arguments it can be concluded that tetrodotoxin (**111**) itself is a *zwitterion*, one of whose hydroxyl groups is deprotonated.

The answer to the next question, which of the hydroxyl groups of tetrodotoxin is acidic enough to lose a proton to be gained by nitrogen, came through analysis of the nmr spectrum of heptaacetylanhydrotetrodotoxin. This derivative was one of a mixture of polyacetates which was formed when tetrodotoxin (**111**) was acetylated for ten days at 25° (Woodward, 1964).

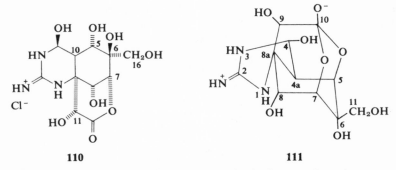

110	**111**

In order to arrive at a structure for heptaacetylanhydrotetrodotoxin, one needs to examine the hypothetical precursor of derivative **109**, i.e., a hydrochloride of tetrodotoxin minus the methyl ether and isopropylidene functions, which is represented by structure **110**. If this compound **110** were to be acetylated,

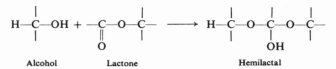

SCHEME 4.4

three carbons of the carbocyclic ring, C-5, C-7, and C-8, would bear single hydrogen atoms as well as acyloxy groups. In fact, the nmr spectrum of the heptaacetylanhydro derivative exhibited only one such resonance. Consequently, two of the three involved hydroxy groups apparently combined to form a new system. If, as is shown in Scheme 4.4, one of the hydroxyl groups combines with the lactone function to form a hemilactal grouping, only one hydroxyl group remains to be acetylated. Double resonance experiments on the heptaacetylanhydro derivative showed that the hydroxyl group at C-8 is being acetylated while the hydroxyl group at C-5 participates in the hemilactal formation (Woodward, 1964). On the basis of this last consideration structure **111** may be written for tetrodotoxin, and **112** for the heptaacetylanhydro derivative.

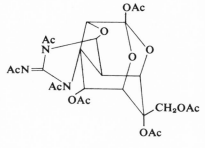

112

In the spring of 1964, when this complex structural problem had been resolved, there remained the possibility, suggested by Professor Tsuda, that tetrodotoxin might not be a C_{11} monomer as represented by **111**, but a C_{22} dimer. This question has since been decided in favor of the monomeric structure. Woodward and Gougoutas (1964) succeeded in preparing crystals of tetrodotoxin that were suitable for single-crystal X-ray diffraction studies. Measurement of the unit cell dimensions and density of the crystals together with a consideration of symmetry requirements led to the unambiguous conclusion that crystalline tetrodotoxin is monomeric and contains two molecules per unit cell. Goto *et al.* (1965) arrived at the same result for tetrodotoxin in solution through a careful analysis of its titration curves.

Although no complete synthesis of tetrodotoxin has been published to date, the toxin, which has become a valuable tool in pharmacological research for sodium ion transport studies across cell membranes, is available from puffer fish ovaries. Tsuda (1966) has described his isolation procedure, which yields 8–9 g of toxin from 1000 kg of puffer fish ovaries. The isolation scheme detailed by Goto *et al.* (1965) furnishes 1–2 g of crystalline toxin from 100 kg of ovaries. Average weight of a puffer ovary is 200 g.

In 1963 Brown and Mosher isolated from the egg clusters of the California newt, *Taricha torosa*, a potent neurotoxin which they named tarichatoxin. It soon became evident that the two toxins from two totally unrelated sources, tetrodotoxin from puffers and tarichatoxin from salamanders, were identical (Buchwald *et al.*, 1964). Mosher and co-workers (1964) brought their structural studies to a successful conclusion at the same time as did the three groups working on tetrodotoxin.

Recently Hashimoto and Noguchi (1971) have reported the isolation of a toxin almost certainly identical with, or closely related to tetrodotoxin from yet another source, a goby, *Gobius criniger*, from the Ryukyu islands. Close similarity or identity of the two toxins was suggested by identical dose-response curves in the bioassay and by identical tlc spots in three solvent systems.

c. Synthetic approaches

Two research groups have so far published details of their efforts toward a total synthesis of tetrodotoxin. One of these is that of Keana at the University of Oregon; the other is that of Goto at Nagoya University, who had participated earlier in the structural work with Professor Hirata (Goto *et al.*, 1965).

Goto's ingenious approach to a total synthesis of tetrodotoxin (**111**) has been outlined in two published communications (Kishi *et al.*, 1970a,b). Perhaps the most remarkable features of Goto's synthesis are the well-chosen starting material and the stereospecificity of each step, which achieve an intermediate that possesses all six centers of chirality of the carbocyclic ring of tetrodotoxin. The synthesis, which is presented in Chart 4.5, proceeds from a Diels-Alder reaction of butadiene (**117**) with 5-(α-oximinoethyl)-toluquinone (**114**) in acetonitrile in the presence of stannic chloride in yield of 83%. Adduct **115a** is converted to mesylate **115b**, which in boiling water undergoes a Beckmann rearrangement to amide **116**. Sodium borohydride in methanol reduces selectively and stereospecifically the carbonyl at C-5 (tetrodotoxin numbering) to yield alcohol **117**, which with *m*-chloroperbenzoic acid and camphorsulfonic acid is transformed to hydroxyether **118**. Alkaline hydrogen peroxide treatment of **118** transformed the remaining olefinic linkage into epoxide **119**. This epoxide on sodium borohydride reduction lost its remaining carbonyl group and was transformed into alcohol **120a**. The corresponding epoxide acetate (**120b**) when treated with a 1:1 mixture of sulfuric and acetic acids at −20° was opened in desired fashion and yielded key intermediate **121**, which possesses the necessary functionality and the correct stereochemistry toward a total synthesis of tetrodotoxin. Correctness of intermediate structures and stereochemistry was established by nmr analysis and by preparation and comparison with several epimeric compounds.

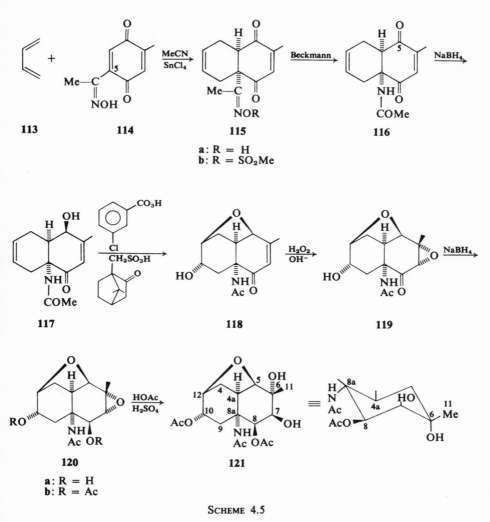

SCHEME 4.5

Goto's successful completion of his total synthesis of tetrodotoxin has now been reported (Goto, 1971).

It may be recalled that a great variety of drastic degradations of tetrodotoxin (**111**) led to a number of closely related quinazoline derivatives **108**. Once the complete structure of tetrodotoxin was known it became evident that the molecule indeed contained a hydroquinazoline skeleton possessing a carbon substituent in one of the angular positions. Keana and co-workers (1969) in their first approach toward a total synthesis of tetrodotoxin achieved the construction of a suitably substituted hydroquinazoline by way of a novel Diels-Alder reaction. The heterocyclic dienophile **124** resulted from

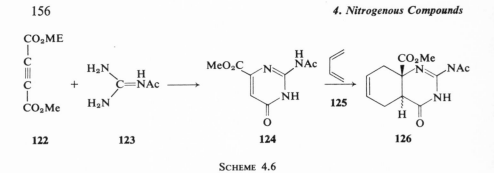

SCHEME 4.6

condensation of methyl acetylene dicarboxylate (122) and acetylguanidine 123. Pyrimidone 124 reacted with butadiene (125) in tetrahydrofuran at 140° for 2 days and yielded quinazolone derivative 126. This is summarized in Scheme 4.6.

Keana's second and totally different approach (Keana and Kim, 1970, 1971) is based on the interesting observation that the important biosynthetic intermediate shikimic acid (127) possesses four of the requisite functions of the carbocyclic ring of tetrodotoxin in addition to a double bond that can serve as a handle to the introduction of further functionality. In their first exploration of this approach Keana and Kim (1970) succeeded in preparing a number of pyrazolines of general structure 128 by reacting various derivatives of shikimic acid (127) with diazomethane.

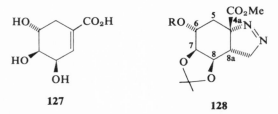

127   128

Keana's subsequent goal (Keana and Kim, 1971), was the lengthening of the carboxyl group at C-4a (tetrodotoxin numbering) of a suitable pyrazoline such as 128 to the necessary two-carbon side chain and connecting it with the oxygen function at C-6, which is a crucial feature of the tetrodotoxin molecule. After many interesting though fruitless attempts Keana and Kim (1971) reached their goal in the key intermediate 131 by chain lengthening of the methyl carboxylate in 128 via its acid chloride and α-ketoacetamide 129a produced by Ugi and Fetzer's (1961) methylisocyanide reaction. From the ketoamide 129a to corresponding mesylate 129b, which by sodium borohydride reduction furnished an alcohol 130, a desired lactone 131 could be obtained by treatment of 130 in refluxing pyridine–water.

Remarkably, the two published routes toward a total tetrodotoxin synthesis though very different in approach and detail arrive at closely parallel intermediates—a carbocyclic ring with functionality and chirality closely resembling tetrodotoxin itself.

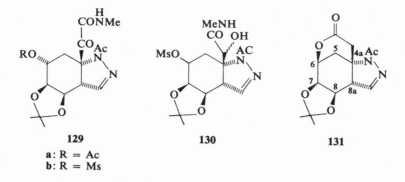

129

130

131

a: R = Ac
b: R = Ms

4. *Cypridina* LUCIFERIN

The complexity of the structure of *Cypridina* luciferin lies primarily in the accumulation of diverse heterocyclic systems and in its instability rather than in a high ratio of hetero to carbon atoms as is the case with tetrodotoxin (111) or saxitoxin (107). The molecule is extended rather than fused and thus presented fewer difficulties in structural elucidation than might have been expected of a molecule having a composition of $C_{22}H_{27}N_7O \cdot 2\,HX$ (Kishi *et al.*, 1966a).

The intriguing phenomenon of bioluminescence has been observed and described by naturalists and biologists in a wide variety of organisms ranging from bacteria to fish for many years. Even now, chemical structures and detailed mechanisms are known for only very few organisms, one of them the bioluminescence of the ostracod crustacean *Cypridina hilgendorfii*, a tiny, so-called seed shrimp, about 3 mm long, that occurs abundantly along the coasts of Japan. Although the compound that is responsible for the bioluminescence is referred to as *Cypridina* luciferin, it is structurally unrelated to the much better known firefly luciferin (Goto and Kishi, 1968).

Structural studies date from the time when crystalline *Cypridina* luciferin was first obtained by Shimomura *et al.* (1957) as orange-red needles. This was followed by recognition of a few of the structural parts as e.g., indole and guanidine that make up the molecule (Hirata *et al.*, 1959). Its bioluminescence mechanism is considered to be relatively simple since the reaction which accompanies light emission requires only the enzyme luciferase and oxygen. The resulting products *Cypridina* oxyluciferin and *C*. etioluciferin can also be produced by reaction with ammonia alone, however without luminescence. It appears that oxyluciferin is the initial product, which with acid may be transformed to etioluciferin and 2-keto-3-methylpentanoic acid (132). The end product etioluciferin is far less sensitive to light and moisture than is *C*. luciferin and it was therefore chosen for structural work (Kishi *et al.*, 1966a).

The simplest of the three compounds, *C*. etioluciferin, has a composition of $C_{16}H_{19}N_7 \cdot 2\,HX$. On the basis of spectral data and chemical degradations its structure was deduced to be **133**, an indole nucleus substituted in the *beta* position by a 2,3-disubstituted pyrazine (Kishi *et al.*, 1966a). This was fully confirmed by total synthesis (Kishi *et al.*, 1966b).

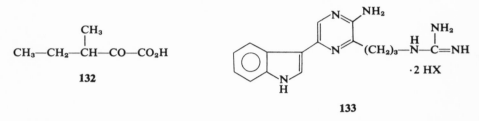

Oxyluciferin, of likely composition $C_{22}H_{27}N_7O_2 \cdot 2\,HX$, can be formally constructed from etioluciferin, $C_{16}H_{19}N_7$, (**133**) and the ketomethylvaleric acid $C_6H_{10}O_3$ (**132**) minus a molecule of water. Its mode of formation and its nmr spectrum led Kishi *et al.* (1966a) to propose structure **134** for *C.* oxyluciferin. Subsequent studies of the mechanism of the chemiluminescence process and the recognition that carbon dioxide is lost during the reaction led Goto and co-workers (1968) to revise the structure of the sensitive *Cypridina* oxyluciferin from **134** to **135**.

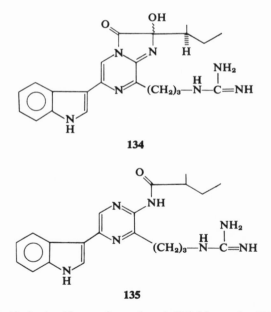

Cypridina luciferin itself was formulated (Kishi *et al.*, 1966a) as **136**, a structure that was confirmed by total synthesis (Kishi *et al.*, 1966b) from *C.* etioluciferin, however, without isolation of pure intermediates and in a yield of less than 1%.

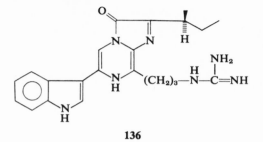

136

It is worth noting that this fascinating and sensitive molecule, *Cypridina* luciferin, is formally made up of the three amino acid or equivalent moieties tryptamine (**137**), arginine (**138**), and isoleucine (**139**), although the three units are linked in unconventional fashion.

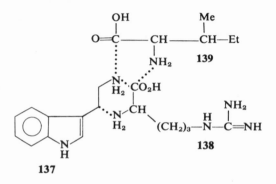

5. CHLOROPHYLL c

None of the compounds that have been discussed so far in this section are closely related to a known natural product from terrestrial sources. In contrast, chlorophyll c is closely related to the principal green plant pigment chlorophyll a (**140**). However, it is a unique variant of chlorophyll a and is an important photosynthetic pigment of marine diatoms, dinoflagellates, and brown algae. It was isolated as a crystalline bistetrahydrofuranate (Dougherty *et al.*, 1966) from the diatom *Nitzschia closterium* and has now been con- clusively demonstrated as a mixture of magnesium tetradehydropheno- porphyrin a_5 (**141a**) monomethyl ester and its hexadehydro analog (**141b**) (Dougherty *et al.*, 1970). Important structural differences between the two chlorophylls occur in ring IV—chlorophyll c possesses a $\Delta^{7,8}$-olefin and an acrylic acid side chain at C-7. Because of the highly specialized nature of chlorophyll chemistry no additional details will be presented here. Instead, the reader is referred to recent reviews and monographs on chlorophyll (*inter alia*, Inhoffen, 1968; Vernon and Seely, 1966).

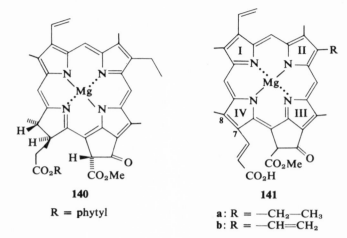

140

R = phytyl

141

a: R = —CH$_2$—CH$_3$
b: R = —CH=CH$_2$

6. SURUGATOXIN

In 1967 Hashimoto and co-workers (Hashimoto *et al.*, 1967b) followed up earlier reports of human intoxications that had been caused by ingestion of a gastropod mollusk *Babylonia japonica* (Japanese ivory shell) and isolated a toxin from the midgut gland of the mollusk. Hashimoto *et al.* (1967) developed a bioassay based on mydriasis in mice since optical disturbances including mydriasis had been reported among the human symptoms of intoxication. Shibota and Hashimoto (1970, 1971) further purified and characterized the toxin.

Another group of Japanese workers (Kosuge *et al.*, 1972) reisolated the toxin they call surugatoxin (after Suruga Bay, where toxic outbreaks had occurred), and determined its structure by X-ray crystallographic techniques. Surugatoxin is a unique molecule, which combines 6-bromoxindole (**143**) with a pteridine (**144**) through a spiro linkage in the β-position of the indole ring. A carboxylic acid group is esterified with myoinositol. The structure of surugatoxin (**142**) bears no resemblance to that of the *Belladonna* alkaloid atropine (**145**), with which it shares mydriatic activity.

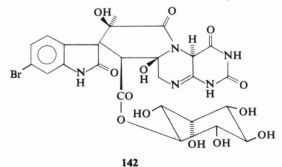

142

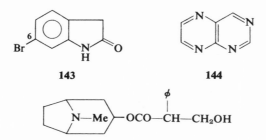

143   **144**

145

TABLE 4.1

<small>Nitrogenous Compounds from Marine Organisms</small>

Text no.	Name	mp (in degrees)	$[\alpha]_D^\circ$	References
18	γ(Guanylureido)butyric acid (gongrine)	208–209 (dec)		Ito and Hashimoto (1965)
21	α-Amino-γ-(guanylureido) valeric acid (gigartinine)	197	+7.5	Ito and Hashimoto (1966a,b)
	—nitrate	(dec)		
22	Trimethyl(2-carboxy-3-hydroxypropyl) ammonium chloride (atrinine hydrochloride)	149–151	0 $[\alpha]_{226}$ +7728 $[\alpha]_{195}$ −9192	Konosu *et al.* (1970)
26	Murexine picrate (choline urocanoate picrate)	221–222 (dec)		Erspamer and Benati (1953b)
29	Choline β,β-dimethyl-acrylate	—		Keyl *et al.* (1957); Whittaker (1959a)
	—aurichloride	97		
31	Choline acrylate	—		Whittaker (1959b)
32	Choline-3-acetoxyhexadecanoate (pahutoxin)	74–75	+3	Boylan and Scheuer (1967)
39	2,6 Dibromo-4-hydroxy-4-acetamidocyclohexa-2,5-dienone	193–195		Sharma and Burkholder (1967b)
40	2,6-Dibromo-4-hydroxy-4-acetamidocyclohexa-2,5-dienone dimethyl ketal	191		Sharma *et al.* (1968)
42a	Aeroplysinin-1	120 112–116	+186 −198	Fattorusso *et al.* (1970a); Fulmor *et al.* (1970)
45	Aerothionin	134–137	+252	Fattorusso *et al.* (1970b)
48	Homoaerothionin	amorph		Fattorusso *et al.* (1971a)
	—diacetate	166–167	+191.5	
51	Caulerpicin	95		Santos and Doty (1968)

(continued)

TABLE 4.1—*continued*

Text no.	Name	mp (in degrees)	$[\alpha]_D°$	References
52	4-Dimethylamino-1,2-dithiolane (nereistoxin)	bp 212–213		Hashimoto and Okaichi (1960)
	—hydrogen oxalate	168–170		
65	6,6′-Dibromoindigotin (Tyrian purple)	subl.		Friedländer (1907)
67	6-Bromo-2-methylmer-captoindoxyl-3-sulfate, silver salt	118–120 (dec)		Baker and Sutherland (1968)
73	2-(2-Hydroxy-3,5-dibromophenyl)-3,4,5-tribromopyrrole	135–155 (dec)		Burkholder *et al.* (1966)
78a	4,5-Dibromopyrrole-2-carboxylic acid	148 subl.	—	Forenza *et al.* (1971)
78b	4,5-Dibromopyrrole-2-carboxylic acid amide	164–166	—	Forenza *et al.* (1971)
78c	4,5-Dibromopyrrole-2-nitrile	172–173	—	Forenza *et al.* (1971)
79	3-Carboxymethyl-4-isopropenyl-pyrroli-dine-2-carboxylic acid (α-kainic, digenic acid)	251 (dec)	−14.8	Murakami *et al.* (1953)
82	3-Carboxymethyl-4-isopropenylpyrrolidine-2-carboxylic acid (α-allokainic acid)	237–238 (dec)	+6.7	Murakami *et al.* (1955b)
84	L$_s$-Arabo-2-carboxy-4-(1-methyl-5-carboxy-*trans,trans,s-trans*-1,3-hexadienyl)-3-pyrroli-dine acetic acid (domic acid)	217 (dec)	−109.6	Daigo (1959)
86	Aplysioviolin	315 (dec)		Rüdiger (1967a)
87	2-(3-Pyridyl)-3,4,5,6-tetrahydropyridine (anabaseine)—picrate	172–175		Kem *et al.* (1971)
89	3,4-Dihydroxyquinoline-2-carboxylic acid	253–254 (dec)		Fattorusso *et al.* (1971)
90	6,13-Dihydrodibenzo[b,i]phenazine-5,12-dicarboxylate (caulerpin)	317		Santos (1970)
91a	Dibromophakellin	237–245 (dec)	−203	Sharma and Burkholder (1971)

(continued)

TABLE 4.1—*continued*

Text no.	Name	mp (in degrees)		References
91b	4-Bromophakellin	170–180 (dec)		Sharma and Burkholder (1971)
92	Oroidin	—	—	Forenza *et al.* (1971)
107	Saxitoxin	—	+133	Schantz *et al.* (1957)
111	Tetrodotoxin	> 300	−8.64	Tsuda and Kawamura (1952, 1953)
136	*Cypridina* luciferin	182–195		Shimomura *et al.* (1957)
141	Chlorophyll c	—		Dougherty *et al.* (1966)
142	Surugatoxin	> 300		Kosuge *et al.* (1972)

REFERENCES

Ackermann, D., and List, P. H. (1958). *Hoppe-Seyler's Z. Physiol. Chem.* **313**, 30.

Ackermann, D., and List, P. H. (1960). *Hoppe-Seyler's Z. Physiol. Chem.* **318**, 281.

Ackermann, D., and Pant, R. (1961). *Naturwissenschaften* **48**, 646.

Asano, M., and Itoh, M. (1960). *Ann. N.Y. Acad. Sci.* **90**, 674.

Bacq, Z. M. (1937). *Arch. Int. Physiol.* **44**, 190.

Baeyer, A. (1879). *Chem. Ber.* **12**, 1309.

Baker, J. T., and Sutherland, M. D. (1968). *Tetrahedron Lett.* 43.

Baslow, M. (1969). "Marine Pharmacology." Williams & Wilkins, Baltimore, Maryland.

Bergquist, P. R., and Hartman, W. D. (1969). *Mar. Biol.* **3**, 247.

Bouchilloux, S., and Roche, J. (1954a). *C. R. Soc. Biol.* **148**, 1583.

Bouchilloux, S., and Roche, J. (1954b). *C. R. Soc. Biol.* **148**, 1732.

Bouchilloux, S., and Roche, J. (1955). *Bull. Inst. Oceanogr.* **52**, 1054.

Boylan, D. B., and Scheuer, P. J. (1967). *Science* **155**, 52.

Brock, V. E. (1955). *Copeia* 195.

Buchwald, H. D., Durham, L., Fischer, H. G., Harada, R., Mosher, H. S., Kao, C. Y., and Fuhrman, F. A. (1964). *Science* **143**, 474.

Burkholder, P. R., Pfister, R. M., and Leitz, F. H. (1966). *Appl. Microbiol.* **14**, 649.

Brown, M. S., and Mosher, H. S. (1963). *Science* **140**, 295.

Chambers, J. R., and Isbell, A. G. (1964). *J. Org. Chem.* **29**, 832.

Chapman, D. J., and Fox, D. L. (1969). *J. Exp. Mar. Biol. Ecol.* **4**, 71.

Chapman, D. J., Cole, W. J., and Siegelman, H. W. (1967). *J. Amer. Chem. Soc.* **89**, 5976.

Ciereszko, L. S., Odense, P. K., and Schmidt, R. W. (1960). *Ann. N.Y. Acad. Sci.* **90**, 920.

Cosulich, D. B., and Lovell, F. M. (1971). *J. Chem. Soc. D* 397.

Daigo, K. (1959). *Yakugaku Zasshi* **79**, 353 [*Chem. Abstr.* **53**, 14218 (1959)].

Doty, M. S., and Santos, G. A. (1966). *Nature (London)* **211**, 990.

Dougherty, R. C., Strain, H. H., Svec, W. A., Uphaus, R. A., and Katz, J. J. (1966). *J. Amer. Chem. Soc.* **88**, 5037.

Dougherty, R. C., Strain, H. H., Svec, W. A., Uphaus, R. A., and Katz, J. J. (1970). *J. Amer. Chem. Soc.* **92**, 2826.

Erspamer, V. (1948). *Experientia* **4**, 226.

Erspamer, V., and Benati, O. (1953a). *Science* **117**, 161.

Erspamer, V., and Benati, O. (1953b). *Biochem. Z.* **324**, 66.

Fattorusso, E., Minale, L., and Sodano, G. (1970a). *Chem. Commun.* 751.

Fattorusso, E., Minale, L., Sodano, G., Moody, K., and Thomson, R. H. (1970b). *Chem. Commun.* 752.

Fattorusso, E., Minale, L., Moody, K., Sodano, G., and Thomson, R. H. (1971a). *Gazz. Chim. Ital.* **101**, 61.

Fattorusso, E., Forenza, S., Minale, L., and Sodano, G. (1971b). *Gazz. Chim. Ital.* **101**, 104.

Fattorusso, E., Minale, L., and Sodano, G. (1972). *J. Chem. Soc. Perkin Trans.* **1**, 16.

Flury, F. (1915). *Arch. Exp. Pathol. Pharmakol.* **79**, 250.

Forenza, S., Minale, L., Riccio, R., and Fattorusso, E. (1971). *J. Chem. Soc. D* 1129.

Friedländer, P. (1907). *Monatsh. Chem.* **28**, 991.

Friedländer, P. (1909). *Chem. Ber.* **42**, 765.

Fulmor, W., Van Lear, G. E., Morton, G. O., and Mills, R. D. (1970). *Tetrahedron Lett.* 4551.

Gasteiger, E. L., Haake, P. C., and Gergen, J. A. (1960). *Ann. N.Y. Acad. Sci.* **90**, 622.

Goto, T. (1971). Natural Products Seminar, Univ. of Hawaii, Honolulu, November.

Goto, T., and Kishi, Y. (1968). *Angew. Chem. Int. Ed. Engl.* **7**, 407.

Goto, T., Kishi, Y., Takahashi, T., and Hirata, Y. (1965). *Tetrahedron* **21**, 2059.

Goto, T., Inoue, S., Sugiura, S., Nishikawa, K., Isobe, M., and Abe, Y. (1968). *Tetrahedron Lett.* 4035.

Hagiwara, H., Numata, M., Konishi, K., and Oka, Y. (1965). *Chem. Pharm. Bull. (Tokyo)* **13**, 253.

Halstead, B. W. (1967). "Poisonous and Venomous Marine Animals of the World," Vol. 2. U.S. Government Printing Office, Washington, D.C., pp. 679 ff.

Hanessian, S., and Kaltenbronn, J. S. (1966). *J. Amer. Chem. Soc.* **88**, 4509.

Hashimoto, Y., and Okaichi, T. (1960). *Ann. N.Y. Acad. Sci.* **90**, 667.

Hashimoto, Y., and Noguchi, T. (1971). *Toxicon* **9**, 79.

Hashimoto, Y., Konosu, S., Yasumoto, T., Inoue, A., and Noguchi, T. (1967a). *Toxicon* **5**, 85.

Hashimoto, Y., Miyazawa, K., Kamiya, H., and Shibota, M. (1967b). *Nippon Suisan Gakkaishi* **33**, 661.

Hashimoto, Y., Okaichi, T., Dang, L. D., and Noguchi, T. (1968). *Nippon Suisan Gakkaishi* **34**, 528.

Hirata, Y., Shimomura, O., and Eguchi, S. (1959). *Tetrahedron Lett.* (5), 4.

Honjo, M., Miyamoto, M., Ueyanagi, J., Nawa, H., and Uchibayashi, M. (1955). *J. Pharm. Soc. Jap.* **75**, 853. [*Chem. Abstr.* **50**, 4120 (1956).]

Hoppe-Seyler, F. A. (1933). *Hoppe-Seyler's Z. Physiol. Chem.* **222**, 105.

Inhoffen, H. H. (1968). *Pure Appl. Chem.* **17**, 443.

Ito, K., and Hashimoto, Y. (1965). *Agr. Biol. Chem.* **9**, 832.

Ito, K., and Hashimoto, Y. (1966a). *Nippon Suisan Gakkaishi* **32**, 274.

Ito, K., and Hashimoto, Y. (1966b). *Nature (London)* **211**, 417.

Ito, K., and Hashimoto, Y. (1969). *Agr. Biol. Chem.* **33**, 237.

Ito, K., Miyazawa, K., and Hashimoto, Y. (1966). *Nippon Suisan Gakkaishi* **32**, 727.

Ito, K., Miyazawa, K., and Hashimoto, Y. (1967). *Nippon Suisan Gakkaishi* **33**, 572.

Jackim, E., and Gentile, J. (1968). *Science* **162**, 915.

Keana, J. F. W., and Kim, C. U. (1970). *J. Org. Chem.* **35**, 1093.

Keana, J. F. W., and Kim, C. U. (1971). *J. Org. Chem.* **36**, 118.

Keana, J. F. W., Mason, F. P., and Bland, J. S. (1969). *J. Org. Chem.* **34**, 3705.

Kem, W. R. (1971). *Toxicon* **9**, 23.

Kem, W. R., Coates, R. M., and Abbott, B. C. (1969). *Fed. Proc. Fed. Amer. Soc. Exp. Biol.* **28**, 610.

Kem, W. R., Abbott, B. C., and Coates, R. M. (1971). *Toxicon* **9**, 15.

Keyl, M. J., Michaelson, I. A., and Whittaker, V. P. (1957). *J. Physiol.* **139**, 434.

Kishi, Y., Goto, T., Hirata, Y., Shimomura, O., and Johnson, F. H. (1966a). *Tetrahedron Lett.* 3427.

Kishi, Y., Goto, T., Inoue, S., Sugiura, S., and Kishimoto, H. (1966b). *Tetrahedron Lett.* 3445.

Kishi, Y., Nakatsubo, F., Aratani, M., Goto, T., Inoue, S., Kakoi, H., and Sugiura, S. (1970a). *Tetrahedron Lett.* 5127.

Kishi, Y., Nakatsubo, F., Aratni, M., Goto, T., Inoue, S., and Kakoi, H. (1970b). *Tetrahedron Lett.* 5129.

Kittredge, J. S., and Hughes, R. R. (1964). *Biochemistry* **3**, 991.

Kittredge, J. S., and Isbell, A. F. (1967). *Biochemistry* **6**, 289.

Kittredge, J. S., Roberts, E., and Simonsen, D. G. (1962). *Biochemistry* **1**, 624.

Konosu, S., Inoue, A., Noguchi, T., and Hashimoto, Y. (1968). *Toxicon* **6**, 113.

Konosu, S., Chen, Y.-N., and Watanabe, K. (1970). *Nippon Suisan Gakkaishi* **36**, 940.

Kosuge, T., Zenda, H., Ochiai, A., Masaki, N., Nogucki, M., Kimura, S., and Narita, H. (1972). *Tetrahedron Lett.* 2545.

Lindberg, B. (1955a). *Acta Chem. Scand.* **9**, 1093.

Lindberg, B. (1955b). *Acta Chem. Scand.* **9**, 1323.

List, P. H. (1958). *Planta Med.* **6**, 424.

Lovell, F. M. (1966). *J. Amer. Chem. Soc.* **88**, 4510.

Madgwick, J. C., Ralph, B. J., Shannon, J. S., and Simes, J. J. H. (1970). *Arch. Biochem. Biophys.* **141**, 766.

Mann, J. A., Jr., and Povich, M. J. (1969). *Toxicol. Appl. Pharmacol.* **14**, 584.

Mold, J. D., Bourden, J. P., Stanger, D. W., Maurer, J. E., Lynch, J. M. Wyler, R. S., Schantz, E. J., and Riegel, B. (1957). *J. Amer. Chem. Soc.* **79**, 5235.

Moody, K., Thomson, R. H., Fattorusso, E., Minale, L., and Sodano, G. (1972). *J. Chem. Soc. Perkin* Trans. **1**, 18.

Morimoto, H. (1955). *J. Pharm. Soc. Jap.* **75**, 943 and papers immediately preceding [*Chem. Abstr.* **50**, 4907 (1956)].

Mosher, H. S., Fuhrman, F. A., Buchwald, H. D., and Fischer, H. G. (1964). *Science* **144**, 1100.

Murakami, S., Takemoto, T., Shimizu, Z., and Daigo, K. (1953). *Jap. J. Pharm. Chem.* **25**, 571 [*Chem. Abstr.* **48**, 4774 (1954)].

Murakami, S., Takemoto, T., Tei, Z., and Daigo, K. (1954). *J. Pharm. Soc. Jap.* **74**, 560. [*Chem. Abstr.* **48**, 12676 (1954).]

Murakami, S., Takemoto, T., Tei, Z., and Daigo, K. (1955a). *J. Pharm. Soc. Jap.* **75**, 866, 869 [*Chem. Abstr.* **50**, 4122, 4123 (1956)].

Murakami, S., Takemoto, T., Tei, Z., Daigo, K., and Takagi, N. (1955b). *J. Pharm. Soc. Jap.* **75**, 1252 [*Chem. Abstr.* **50**, 4123 (1952)].

Murakami, S., Takemoto, T., Tei, Z., and Daigo, K. (1955c). *J. Pharm. Soc. Jap.* **75**, 1255 [*Chem. Abstr.* **50**, 4124 (1956)].

Murtha, E. F. (1960). *Ann. N.Y. Acad. Sci.* **90**, 820.

Nakazawa, Y. (1959). *J. Biochem. (Tokyo)* **46**, 1579.

Nawa, H., Nakamori, R., and Matsuoka, T. (1955a). *J. Pharm. Soc. Jap.* **75**, 850 [*Chem. Abstr.* **50**, 4120 (1956)].

Nawa, H., Ueyanagi, J., Nakamori, R., Matsuoka, T., and Kimata, S. (1955b). *J. Pharm. Soc. Jap.* **75**, 860 [*Chem. Abstr.* **50**, 4122 (1956)].

Needler, A. B. (1949). *J. Fish. Res. Bd. Can.* **7**, 490.

Noguchi, T., Konosu, S., and Hashimoto, Y. (1969). *Toxicon* **7**, 325.

Okaichi, T., and Hashimoto, Y. (1962a). *Agr. Biol. Chem.* **26**, 224.

Okaichi, T., and Hashimoto, Y. (1962b). *Nippon Suisan Gakkaishi* **28**, 930.

Pasini, C., Vercellone, A., and Erspamer, V. (1952). *Justus Liebigs Ann. Chem.* **578**, 6.

Rapoport, H., Brown, M. S., Oesterlin, R., and Schuett, W. (1964). *Abstracts*, 147th National Meeting, Amer. Chem. Soc., Philadelphia, p. 3N.

Roberts, E., and Kittredge, J. S. (1969). "Free Amino Acids and Related Substances in Marine Organisms." U.S. Department of Commerce, Document AD 697 976.

Robinson, J. P., Jr. (1971). *Sea Frontiers* **17**, 76.

Rüdiger, W. (1967a). *Hoppe-Seyler's Z. Physiol. Chem.* **348**, 129.

Rüdiger, W. (1967b). *Hoppe-Seyler's Z. Physiol. Chem.* **348**, 1554.

Rüdiger, W., Klose, W., Tursch, B., Houvenaghel-Crevecoeur, N., and Budzikiewicz, H. (1968). *Justus Liebigs Ann. Chem.* **713**, 209.

Russell, F. E. (1967). *Fed. Proc. Fed. Amer. Soc. Exp. Biol.* **26**, 1206.

Sachs, F., and Kempf, R. (1903). *Chem. Ber.* **36**, 3299.

Sachs, F., and Sichel, E. (1904). *Chem. Ber.* **37**, 1861.

Santos, G. A. (1970). *J. Chem. Soc. C* 842.

Santos, G. A., and Doty, M. S. (1968). *In* "Drugs from the Sea" (H. D. Freudenthal, ed.), p. 173. Marine Technology Society, Washington, D.C.

Schantz, E. J. (1960). *Ann. N.Y. Acad. Sci.* **90**, 843.

Schantz, E. J. (1969). *J. Agr. Food Chem.* **17**, 413.

Schantz, E. J. (1971). *In* "Microbial Toxins" (S. Kadis, A. Ciegler, and S. J. Ajl, eds.), Vol. 7, pp. 3–26. Academic Press, New York.

Schantz, E. J., and Magnusson, H. W. (1964). *J. Protozool.* **11**, 239.

Schantz, E. J., Mold, J. D., Stanger, D. W., Shavel, J., Riel, F. J., Bowden, J. P., Lynch, J. M., Wyler, R. S., Riegel, B., and Sommer, H. (1957). *J. Amer. Chem. Soc.* **79**, 5230.

Schantz, E. J., Mold, J. D., Howard, W. L., Bowden, J. P., Stanger, D. W., Lynch, J. M., Wintersteiner, D. P., Dutcher, J. D., Walters, D. R., and Riegel, B. (1961). *Can. J. Chem.* **39**, 2117.

Schantz, E. J., Lynch, J. M., Vayvada, G., Matsumoto, K., and Rapoport, H. (1966). *Biochemistry* **5**, 1191.

Scheuer, P. J. (1964). *Fortschr. Chem. Org. Naturst.* **22**, 265.

Schuett, W., and Rapoport, H. (1962). *J. Amer. Chem. Soc.* **84**, 2266.

Sharma, G. M., and Burkholder, P. R. (1967a). *J. Antibiot. (Tokyo) Ser. A* **20**, 200.

Sharma, G. M., and Burkholder, P. R. (1967b). *Tetrahedron Lett.* 4147.

Sharma, G. M., and Burkholder, P. R. (1971). *Chem. Commun.* 151.

Sharma, G. M., Vig, B., and Burkholder, P. R. (1968). *In* "Drugs from the Sea" (H. D. Freudenthal, ed.), p. 119. Marine Technology Society, Washington, D.C.

Sharma, G. M., Vig, B., and Burkholder, P. R. (1970). *J. Org. Chem.* **35**, 2823.

Shibota, M., and Hashimoto, Y. (1970). *Nippon Suisan Gakkaishi* **36**, 115.

Shibota, M., and Hashimoto, Y. (1971). *Nippon Suisan Gakkaishi* **37**, 936.

Shimomura, O., Goto, T., and Hirata, Y. (1957). *Bull. Chem. Soc. Jap.* **30**, 929.

Sommer, H., and Meyer, K. F. (1937). *Arch. Pathol.* **24**, 560.

Sommer, H., Whedon, W. F., Koford, C. A., and Stohler, R. (1937). *Arch. Pathol.* **24**, 537.

Späth, E., and Mamoli, L. (1936). *Chem. Ber.* **69**, 1082.

Steiner, M., and Hartmann, T. (1968). *Planta* **79**, 113.

Stempien, M. F., Jr. (1966). *Amer. Zool.* **6**, 363.

Sugawa, T., Sanno, Y., and Kurita, A. (1955a). *J. Pharm. Soc. Jap.* **75**, 845 [*Chem. Abstr.* **50**, 4119 (1956)].

Sugawa, T., Sanno, Y., and Kurita, A. (1955b). *J. Pharm. Soc. Japan* **75**, 856 [*Chem. Abstr.* **50**, 4121 (1956)].

Takemoto, T., Daigo, K., Kondo, Y., and Kondo, K. (1966). *Yakugaku Zasshi* **86**, 874 [*Chem. Abstr.* **66**, 28604 (1967)].

Tawara, Y. (1909). *Yakugaku Zasshi* **29**, 587.

Thomson, D. A. (1963). Ph.D. Dissertation, University of Hawaii, Honolulu.

Thomson, D. A. (1964). *Science* **146**, 244.

Tsuda, K. (1966). *Naturwissenschaften* **53**, 171.

Tsuda, K., and Kawamura, M. (1952). *J. Pharm. Soc. Japan* **72**, 711 [*Chem. Abstr.* **46**, 9733 (1952)].

Tsuda, K., and Kawamura, M. (1953). *Pharm. Bull.* **1**, 112.

Tsuda, K., Ikuma, S., Kawamura, M., Tachikawa, R., Baba, Y., and Miyadera, T., (1962a). *Chem. Pharm. Bull. (Tokyo)* **10**, 856.

Tsuda, K., Ikuma, S., Kawamura, M., Tachikawa, R., and Miyadera, T. (1962b). *Chem. Pharm. Bull. (Tokyo)* **10**, 865.

Tsuda, K., Ikuma, S., Kawamura, M., Tachikawa, R., and Miyadera, T. (1962c). *Chem. Pharm. Bull. (Tokyo)* **10**, 868.

Tsuda, K., Ikuma, S., Kawamura, M., Tachikawa, R., Sakai, K., Tamura, C., and Amakasu, O. (1964). *Chem. Pharm. Bull. (Tokyo)* **12**, 1357.

Ueno, Y., Nawa, H., Ueyanagi, J., Morimoto, H., Nakamori, R., and Matsuoka, T. (1955). *J. Pharm. Soc. Japan* **75**, 807 and papers immediately following [*Chem. Abstr.* **50**, 4115 (1956)].

Ugi, I., and Fetzer, U. (1961). *Chem. Ber.* **94**, 1116.

Vernon, L. L., and Seely, G. R., Eds. (1966). "The Chlorophylls." Academic Press, New York.

Welsh, J. H., and Prock, P. B. (1958). *Biol. Bull.* **115**, 551.

Whittaker, V. P. (1959a). *Biochem. J.* **71**, 32.

Whittaker, V. P. (1959b). *Biochem. Pharmacol.* **1**, 342.

Whittaker, V. P. (1960). *Ann. N.Y. Acad. Sci.* **90**, 695.

With, T. K. (1968). "Bile Pigments." Academic Press, New York.

Wong, J. L., Brown, M. S., Matsumoto, K., Oesterlin, R., and Rapoport, H. (1971a). *J. Amer. Chem. Soc.* **93**, 4633.

Wong, J. L., Oesterlin, R., and Rapoport, H. (1971b). *J. Amer. Chem. Soc.* **93**, 7344.

Woodward, R. B. (1964). *Pure Appl. Chem.* **9**, 49.

Woodward, R. B., and Gougoutas, J. Z. (1964). *J. Amer. Chem. Soc.* **86**, 5030.

Yokoo, A. (1948). *Bull. Tokyo Inst. Technol.* **13**, 8 [*Chem. Abstr.* **44**, 3622 (1950)].

5

NONAROMATIC COMPOUNDS WITH
UNBRANCHED CARBON SKELETONS

As has been pointed out in the Preface, the group of compounds to be discussed in this chapter is perhaps the least coherent and therefore the most controversial in the book. Although few chemists will argue whether a given natural product should be classified as a steroid or an alkaloid, it can become a matter of dispute whether to include branched fatty acids and their derivatives (the so-called iso acids for example) among the compounds possessing "unbranched carbon skeletons." (On the other hand, unbranched benzenoid compounds have been deliberately omitted from discussion here and are being treated in a separate chapter.) Presumably, these branched compounds have arisen biogenetically from an occasional propionate residue being condensed with a preponderance of acetate fragments. All of these compounds are included in this chapter provided they are distinctly *not* of isoprenoid origin, nor have aromatic character. Generally, a positive criterion would be preferable, but no such simple designation embracing these compounds occurs to me.

A. Fatty Acids

There are a number of reasons why the chemistry of fatty acids will receive only summary mention rather than detailed discussion. Most importantly, unmodified fatty acids are not natural products in the historical sense, i.e., they are primary rather than secondary metabolites. Furthermore, it is too soon to know whether marine plants and animals elaborate any unique fatty acids. While fatty acid research that is related to marine organisms, prominently of course to fishes and mammals, dates back many years (for reviews see Hilditch and Williams, 1964; Stansky, 1967), much of the work prior to about 1960 suffers from lack of adequate separation techniques (glc) and of physical tools of structural determination (particularly mass spectrometry). Additionally, the bulk of all the early work on fatty acid chemistry was undertaken with a nutritional (percent saturated vs. unsaturated fatty acids) rather than a structural bias, thus rendering the results only marginally relevant to our discussion. Other interesting fatty acid research has been undertaken primarily from other than a structural chemical viewpoint. For example, the apparently rare occurrence of polyunsaturated acids in blue-green algae (Holton et al., 1968) is essentially a chemotaxonomic investigation; or the stereochemistry of fatty acid dehydrogenation (Morris et al., 1968) was undertaken in connection with biosynthetic work; or, e.g., the work on fatty acid composition of blue-green algae by Parker et al. (1967) was part of a geochemical study. All these researches are only of peripheral interest to our theme. Finally, the sum total of modern and rigorous structural research on marine-derived fatty acids is still quite modest and our ability to draw general conclusions is therefore limited.

From the most recent comprehensive reviews on marine lipids (Lovern, 1964; Malins and Wekell, 1970) it appears that the fatty acid make-up of marine lipids is complex and that its composition in a particular organism may well be characteristic of a given species of plant or animal. It remains to be seen, however, whether any uniquely marine structural features will eventually emerge, e.g., 6,9,12,15-hexadecatetraenoic acid of the marine diatom *Biddulphia sinensis* with its unusual terminal double bond (Klenk and Eberhagen, 1962); or the twenty-two carbon acid with six carbon–carbon double bonds (docosahexaenoic acid), at C-4, C-10, C-13, C-16, and C-19, which has been identified from the dinoflegellate *Gyrodinium cohnii* (Harrington and Holz, 1968) and from cod liver oil (Hinchcliffe and Riley, 1971). It would, for instance, be not at all surprising if some unique halogenated fatty acids were to be isolated from marine sources.

B. Hydrocarbons

Hydrocarbons with an unbranched carbon skeleton have an evident biogenetic relationship to fatty acids. Since in the course of biosynthesis at least one transformation, i.e., decarboxylation, must take place, an opportunity arises for concomitant and independent, or for decarboxylation-triggered reactions and therefore a possibility for the generation of compounds of intrinsic chemical or biological interest.

Not all hydrocarbons that are isolated from marine organisms need necessarily be derived via biosynthesis. The possibility of isolating substances of extraneous origin that have resulted from spillage or dumping of oil or from other man-made pollutants have become increasingly apparent in recent years. Boylan and Tripp (1971) have demonstrated that a considerable range of hydrocarbons, notably benzene and naphthalene derivatives, are extractable into sea water from crude oil or kerosene. No doubt these compounds find their way into the marine biosphere and are being accumulated by some organisms.

Because of a traditional association of hydrocarbons with petroleum products, one would expect the concentration of hydrocarbons in plants or animals to be rather low. Although this is generally true some notable exceptions have been reported in the literature. Perhaps the most extreme concentration of a hydrocarbon (albeit an isoprenoid one) in a marine source has been found by Heller *et al.* (1957). In a survey of the liver oils of eighteen species of sharks these workers showed that sixteen species contained less than one percent of squalene ($C_{30}H_{50}$), while the liver oil of the shark *Dalatias licha* consisted of 70%, and the liver oil of the shark *Centrophorus uyata* of no less than 90% squalene. Doubtless more typical is the investigation by Lambertsen and Holman (1963) of the hydrocarbon content and composition of herring oil. This oil contained 0.05% of a mixture of hydrocarbons ranging from C_{14} to C_{33}, with odd-numbered compounds predominating, in addition to a number of isoprenoid hydrocarbons. Predominance of odd-numbered compounds is a biogenetic indicator for their origin from even-numbered fatty acids.

As is true for fatty acid research, much of the work on hydrocarbon constituents of marine organisms has been primarily undertaken with an aim other than structural chemical, in this case often geochemical or phylogenetic. Some examples of recent research include the hydrocarbon composition of some blue-green algae, including several marine species (Winters *et al.*, 1969) and the most extensive survey to-date of twenty-four species of green, red, and brown marine algae (Youngblood *et al.*, 1971). These workers identified several new hydrocarbons, a C_{16} compound containing a cyclopropane group (tentatively) from the green algae *Ulva lactuca* and *Enteromorpha compressa* and a number of C_{17}, C_{19}, and C_{21} mono to hexaolefins, none conjugated. The authors (Youngblood *et al.*, 1971) made the interesting observation that in one alga, *Ascophyllum nodosum* the polyunsaturated hydrocarbons occurred exclusively in the reproductive tissue of the plant. The C_{21} hexaene, all *cis*-3,6,8,12,15,18-heneicosahexaene, has also tentatively been identified by Lee *et al.* (1970) from the marine diatom *Skeletonema costatum*. Lee and Loeblich (1971) have since investigated the distribution of the C_{21} hexaene in marine and freshwater algae, as well as the occurrence of its obvious precursor, the corresponding C_{22} hexaenoic acid. In this kind of research, where emphasis is placed on comparative aspects and where many of the individual compounds are well characterized, few compounds need to be isolated. Identification can generally be carried out by gas chromatographic separation and mass spectral analysis of standard and unknown mixtures. Youngblood *et al.* (1971) used ingenious micro-techniques of hydrogenation and ozonization to elucidate the structures of the new olefins.

Moore and his group (Moore *et al.*, 1968) on the other hand, in their investigation of the essential oil of the brown algae *Dictyopteris plagiogramma* and *D. australis* have determined the structures of several new hydrocarbons by actual isolation, chemical degradation, and spectral methods. In interesting contrast to Professor Irie's group in Japan (Irie *et al.*, 1964; Kurosawa *et al.*, 1966) who isolated sesquiterpenoid constituents from the brown alga *Dictyopteris divaricata*, Moore and co-workers (1968) have found the essential oil of *D. plagiogramma* and *D. australis* devoid of terpenoid compounds. The first constituent to be isolated and its structure determined was dictyopterene

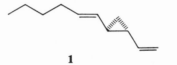

1

A (**1**), which was shown to be *trans*-1-(*trans*-1-hexenyl)-2-vinylcyclopropane. Dictyopterene A was isolated by distillation followed by gas–liquid chromatography. The structure (**1**) was deduced by spectral methods and rigorously confirmed by oxidative degradation to formaldehyde, *n*-valeraldehyde, *trans*-1,2-cyclopropanedicarboxaldehyde, and the corresponding acids. Three groups (Ohloff and Pickenhagen, 1969; Das and Weinstein, 1969; Burgstahler and Groginsky, 1969) have since synthesized racemic dictyopterene A, which is not only of intrinsic chemical interest as a unique C-11 hydrocarbon containing a cyclopropane ring, but which, along with other *Dictyopteris* constituents has a characteristic odor that is generally associated with ocean beaches.

From the same two algal species Pettus and Moore (1970) isolated two additional C_{11} hydrocarbons. One of them was shown to have structure **2**, *trans,cis,cis*-undeca-1,3,5,8-tetraene, by careful analysis of its nmr, uv, and mass spectra.

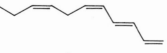

2

The third *Dictyopteris* constituent, present as the major component (50%) of the essential oil of the alga, was designated dictyopterene B and was shown to be *trans*-1(*trans,cis*-hexa-1′,3′-dienyl)-2-vinylcyclopropane (**3**) by nmr spectral analysis and by oxidative degradation to the cyclopropanedicarboxylic acid (Pettus and Moore, 1970). Racemic dictyopterene B has been synthesized by Weinstein's group (Ali *et al.*, 1971).

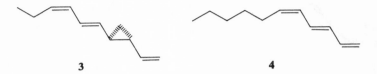

3   **4**

Three additional C_{11} polyolefins, *trans,cis*-undeca-1,3,5-triene (**4**), *trans, trans*-undeca 1,3,5-triene (**5**), and *trans,trans,cis*-undeca-1,3,5,8-tetraene (**6**), have since been isolated from *Dictyopteris*, but experimental details are not yet in print (Pettus and Moore, 1971a).

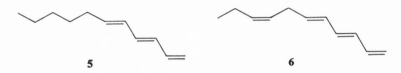

5   **6**

Two further C-11 hydrocarbons, substituted cycloheptadienes, were isolated from *Dictyopteris* by Pettus and Moore (1971b) and were designated dictyopterene C′ (**7**) and D′ (**9**). Dictyopterene C′ (**7**) was shown to be (−)-(R)-6-butylcyclohepta-1,4-diene by direct comparison (except for rotation) with its enantiomer (**8**), derived from dictyopterene A (**1**) by thermal Cope rearrangement (Ohloff and Pickenhagen, 1969; Das and Weinstein, 1969).

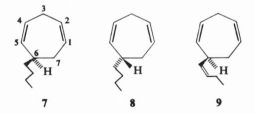

7   **8**   **9**

The configuration of dictyopterene D′ (**9**) was demonstrated to be (+)-6-(*cis*-but-1′-enyl)cyclohepta-1,4-diene by its identity (except for optical properties) with the thermal Cope rearrangement product of dictyopterene B (**3**). Although cycloheptadienes **7** and **9** have been produced *in vitro* from dictyopterenes A and B, their generation *in vivo* presumably must occur from a suitable precursor under mild conditions. Ohloff and Pickenhagen

(1969), e.g., have shown that compound **10** isomerized to dictyopterene
C′ (**7**) at 15°.

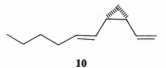

10

In a fascinating parallel investigation Müller (1968) was able to isolate the
sex attractant produced by the female gametes of the brown alga *Ectocarpus
siliculosus* and could show that this chemotactic substance was volatile and
possessed a characteristic odor. By mass culturing the alga (14,900 culture
dishes) over two years Müller *et al.* (1971) were able to isolate 92 mg of
chromatographically pure biologically active substance and to demonstrate
its structure to be 6-(*cis*-but-1′-enyl)cyclohepta-1,4-diene (**11**). The German
workers (Jaenicke *et al.*, 1971) have since synthesized racemic dictyopterene
D′ (**11**) as has the Weinstein group (Ali *et al.*, 1971) by rearranging dictyopterene B (**3**). The full structural elucidation of a sex attractant in a lower
plant is the first for a marine plant and only the second such compound from
any environment. The first compound, sirenin (**12**), was isolated by Machlis
et al. (1966) from the fresh water mold of the genus *Allomyces* and its structure was elucidated by Machlis *et al.* (1968). In view of this precedent Müller

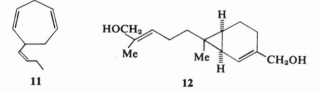

11   **12**

et al. (1971) suggest that the algal sex attractant (**11**) may be referred to as
Ectocarpus sirenin. Regardless of whether plant pheromones will in the future
be referred to by the generic name sirenin, this research will doubtless be
followed by structural elucidation of other chemotactic principles. It will
be recalled (*vide supra*) that Youngblood *et al.* (1971) observed that in the
alga *Ascophyllum nodosum* the polyunsaturated hydrocarbons occur exclusively in the reproductive tissue. Attempts to define chemically a gamone
of a marine alga in fact go back to some work by Cook *et al.* (1948, 1951) who
suggested that the sex attractant material released by the mature eggs of the
brown algae *Fucus vesiculosus* and *F. serratus* may be a C-6 hydrocarbon.
Müller (1972) has reisolated the attractant released by the eggs of *F. serratus*
and has found that it is a small lipophilic molecule with a volatility between
C_8 and C_9 alkanes on a silicone column. Hlubucek *et al.* (1970) have recently
confirmed that *n*-hexane is the major peak (*ca* 60 percent) of the volatiles
extracted from the ripe female tips of *F. vesiculosus*.

C. Miscellaneous Functions

1. SULFUR COMPOUNDS

Moore and his group, in addition to examining the essential oil constituents of two species of brown algae of the genus *Dictyopteris* (Moore *et al.*, 1968; Pettus and Moore, 1970; 1971a,b), have investigated the nonvolatile lipids of these algae. Solvent extraction, followed by column chromatography and gel filtration resulted in the isolation of a series of interesting sulfur-containing compounds, obviously biogenetically related to the dictyopterenes (Roller *et al.*, 1971). The structures of these compounds were determined largely by unambiguous spectral analysis and proved to be *S*-(3-oxoundecyl)-thioacetate (**13**), bis(3-oxoundecyl)disulfide (**14**), *S*-(*trans*-3-oxoundec-4-enyl)-thioacetate (**15**), and 3-hexyl-4,5-dithiacycloheptanone (**16**).

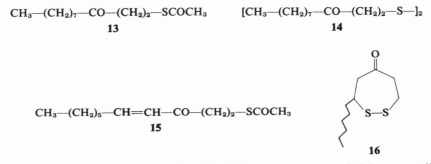

$$CH_3-(CH_2)_7-CO-(CH_2)_2-SCOCH_3$$
13

$$[CH_3-(CH_2)_7-CO-(CH_2)_2-S-]_2$$
14

$$CH_3-(CH_2)_5-CH=CH-CO-(CH_2)_2-SCOCH_3$$
15

16

Compound **14**, bis(3-oxoundecyl)disulfide, was accompanied by a small amount of another sulfur compound, which Moore and co-workers (1972) have now identified as (−)bis(3-acetoxyundec-5-enyl) disulfide (**17**) by spectral means. The mixture of **14** and **17** could be separated only after reductive acetylation to a mixture of *S*-(3-oxoundecyl)thioaceatate (**18**) and *S*(−)(3-acetoxyundec-5-enyl) thioacetate (**19**), which was separable by gel filtration

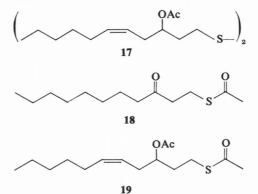

17

18

19

and TLC. Compound **17** could be regenerated from **19** by treatment of **19** with neutral alumina in hexane. Compound **19** additionally is a natural constituent of *Dictyopteris* algae.

In another publication in this series Moore (1971) reported the isolation and structural identification of polysulfides **20** and **21** from the same two species of *Dictyopteris*. The structures of **20**, bis(3-oxoundecyl) trisulfide, and of **21**, bis(3-oxoundecyl)tetrasulfide, were secured by spectral data and by their relationship to the other sulfur-containing *Dictyopteris* constituents. Moore (1971) suggests that the polysulfides may be biogenetic precursors of the dictyopterenes.

$$[CH_3\text{—}(CH_2)_7\text{—}CO\text{—}(C]_2)_2\text{—}S\text{—}H_2S \qquad\qquad [CH_3\text{—}(CH_2)_7\text{—}CO(CH_2)_2\text{—}S\text{—}]_2S_2$$

20 **21**

2. CYCLIC ETHERS

The red algal genus *Laurencia* elaborates, in addition to a variety of sesquiterpenes (see Chapter 1), natural products possessing unbranched carbon chains. Irie and co-workers (1965) isolated from *L. glandulifera* the first of these compounds, which they named laurencin. On the basis of spectral data, hydrogenation to an octahydro derivative, metal hydride reduction of the perhydro compound, hydrolysis of the acetate, but particularly on the basis of detailed nmr analysis including many decoupling experiments, Irie *et al.* (1965) proposed structure **22**, initially without stereochemical designa-

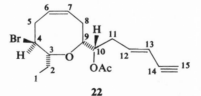

22

tions. The full paper (Irie *et al.*, 1968a) provides additional mass spectral evidence. All structural (including stereochemical) assignments were confirmed by single-crystal X-ray studies (Cameron *et al.*, 1965, 1969). Carbon atom 10 of laurencin was shown to have the *R*-configuration by degradation of octahydrodeacetyllaurencin to (−) atrolactic (2-hydroxy-2-phenylpropionic) acid (**23**) (Irie *et al.*, 1968a).

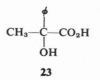

23

A related compound, laureatin (**24**), was isolated by Irie's group (Irie *et al.*, 1968b) from *Laurencia nipponica*. While laureatin (**24**) shares with laurencin

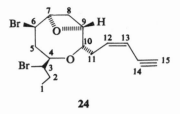

24

(**22**) such structural features as an oxocin ring and a conjugated enyne tail formed from a C-15 unbranched skeleton, it possesses such unique features as an oxetane ring and two bromine substituents. Its structure was secured by reactions paralleling those of laurencin (**22**), by careful and detailed nmr determinations, and by the key degradation of hexahydrolaureatin (zinc and acetic acid, followed by dilute base) to glycol **25**, which was further transformed to the saturated analog **26**. When the acetonide of glycol **26** was

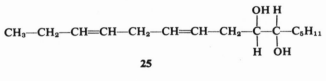

25

$$CH_3-(CH_2)_7-\overset{\overset{\displaystyle OH}{|}}{\underset{\underset{\displaystyle H}{|}}{C}}-\overset{\overset{\displaystyle H}{|}}{\underset{\underset{\displaystyle OH}{|}}{C}}-C_5H_{11}$$

26

compared with a similarly derived degradation product of laurencin (**22**), the spectra of the two sets of derivatives (glycol and acetonide) were superimposable, but they possessed opposite chiralities. This is one of several cases of marine-derived metabolites, where optical antipodes have been isolated from the same or closely related species of plant or animal.

An isomer of laureatin (**24**), designated isolaureatin (**27**) was isolated by Irie and his group (Irie *et al.*, 1968c) from the same red alga, *Laurencia nipponica*. Isolaureatin (**27**) differs from laureatin (**24**) by having a tetrahydrofuran instead of an oxetane ring and by bearing its bromine substituents in positions 3 and 7. Degradative experiments and spectral data established structure **27**, which was further confirmed by comparison of the oxocin acetate **28** with the identical compound derived from laureatin (**24**). The full paper (Irie *et al.*, 1970) confirms all structural assignments for laureatin (**24**) and isolaureatin (**27**) and proposes stereochemical assignments as shown. A likely biogenetic precursor of these *Laurencia* constituents would be hexadeca-4,7,10,13-tetraenoic acid (Irie *et al.*, 1970). The acid-catalyzed rearrangement of laureatin (**24**) to isolaureatin (**27**) has now been discussed in detail (Fukuzawa *et al.*, 1972).

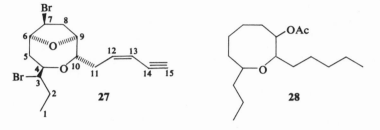

3. LACTONES

Three related lactones, whose skeletons are derived from C_{14} linear dicarboxylic acids that have propyl substituents *alpha* to the two carboxyl groups have been reported from gorgonians (phylum Coelenterata) by the Oklahoma group. Isolation of the first compound, designated ancepsenolide, was first indicated from *Pterogorgia* (syn. *Xiphigorgia*) *anceps* by Ciereszko *et al.* (1960). Its structure, **29**, was elucidated by Schmitz *et al.* (1966) with the aid of spectral data and by a series of degradative experiments. The leading reactions among these were saponification of ancepsenolide (**29**) furnishing the dioxodiacid **30**; ozonolysis of **29** yielding tetradecane-1,14-dioic acid (**31**); and reconversion of the dioxodiacid **30** to ancepsenolide (**29**).

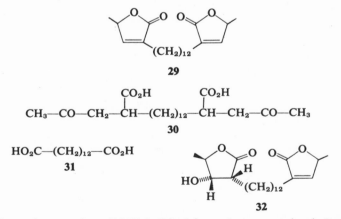

Schmitz and co-workers (1969) isolated from the same animal, *Pterogorgia anceps*, a hydroxyancepsenolide (**32**), convertible to ancepsenolide (**29**) with phosphorus oxychloride in pyridine. Relative stereochemical assignments of the hydroxylated ring as shown in **32** are based on nmr coupling constant data.

A third compound in this series was isolated by Schmitz and Lorance (1971) from the sea fan *Pterogorgia quadalupensis*. This animal yielded ancepsenolide (**29**) in 3.1% yield and the new lactone in 1% yield. The new lactone was shown to be 2-(13'-carboxy-14',15'-diacetoxyhexadecanyl)-2-penten-4-olide (**33**) by spectral data and by conversion to the known ancepsenolide (**29**) by acidic methanolysis followed by treatment with phosphorus oxychloride in pyridine. Relative stereochemistry as shown in **33** was deduced on coupling constant data.

Examination of a third species of *Pterogorgia*, *P. citrina*, surprisingly showed the absence of these or related lactones (Schmitz *et al.*, 1970).

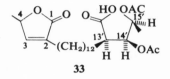

33

4. Keto Acids

Isolation of an epimer of a known substance is by itself not very remarkable and ordinarily would not receive much attention in a context that emphasizes structural chemistry. However, the recent isolation of two prostaglandins by Weinheimer and Spraggins (1969, 1970) is rather remarkable from several viewpoints. The prostaglandins are a group of fatty acid-derived substances that were first isolated from seminal fluid and genital glands of man and sheep, where they occur in trace concentrations. The compounds are characterized by dramatic physiological activity, particularly lowering of blood pressure and stimulating of smooth muscle. Much academic and industrial research effort has been devoted to the biological evaluation and to total synthesis of these compounds. In the light of this background it is indeed fascinating to learn of Weinheimer and Spraggins' (1969, 1970) isolation of two new prostaglandins from a primitive animal, the gorgonian *Plexaura homomalla*, and in yields of 0.2 and 1.3%. Their structures were shown to be **34a** and **b**, and they are designated at 15-*epi*-PGA$_2$ (**34a**) and as 15-*epi*-PGA$_2$ acetate methyl ester (**34b**) since they are epimeric at C-15 with prostaglandin PGA$_2$.

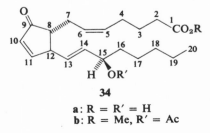

34

a: R = R' = H
b: R = Me, R' = Ac

This stereochemical feature was conclusively demonstrated by ozonolysis of **34b** to methyl hydrogen glutarate (**35**), and 2-acetoxyheptanoic acid (**36a**) which was hydrolyzed to the corresponding $(-)$-2-hydroxyheptanoic acid (**36b**) of known absolute configuration (R).

$$
\begin{array}{c}
CO_2Me \\
(CH_2)_3 \\
CO_2H
\end{array}
$$

35

$$
\begin{array}{c}
OR \\
| \\
HO_2C-CH-(CH_2)_4-CH_3
\end{array}
$$

36

a: R = Ac
b: R = H

The epimeric gorgonian-derived prostaglandins (**34a,b**) do not possess the blood pressure lowering property of their mammalian counterparts. It is not known what if any physiological effects the new prostaglandin isomers exert on coelenterates, members of the phylum that produces them. Schneider and co-workers (1972) have further examined the occurrence of prostaglandins in the gorgonian *P. homomalla* and have found that some specimens of this coelenterate do elaborate prostaglandins that possess the active 15*S* configuration and others elaborate both *R* and *S* constituents. Furthermore the Upjohn group (Bundy *et al.*, 1972a) has isolated from the same gorgonian a new prostaglandin (15*S*)-15-hydroxy-9-oxo-5-*trans*, 10,13-*trans*-prostatrienoic acid (5-*trans*-PGA₂), **37**. Bundy *et al.* (1972b) have also accomplished a laboratory transformation of the 15*R* to the 15*S* configuration.

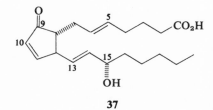

37

TABLE 5.1

Compounds with Unbranched Carbon Skeletons

Text no.	Name	mp (in degrees)	$[\alpha]_D^\circ$	References
1	*trans*-1-(*trans*-1-Hexenyl)-2-vinylcyclopropane (dictyopterene A)	Oil	$+77 \pm 5$	Moore *et al.* (1968)
2	*trans,cis,cis*-Undeca-1,3, 5,8-tetraene	Oil	—	Pettus and Moore (1970)

(continued)

TABLE 5.1—*continued*

Text no.	Name	mp (in degrees)	$[\alpha]_D{}^\circ$	References
3	*trans*-1-(*trans,cis*-Hexa-1′,3′-dienyl)-2-vinyl-cyclopropane (dictyopterene B)	Oil	−43	Pettus and Moore (1970)
7	(−)-(*R*)-6-Butylcyclo-hepta-1,4-diene (dictyopterene C′)	Oil	−13	Pettus and Moore (1971b)
9	(+)-6-(*cis*-But-1′-enyl) cyclohepta-1,4-diene (dictyopterene D′)	Oil	—	Pettus and Moore (1971b)
13	*S*-(3-Oxoundecyl)thio-acetate	Oil	—	Roller *et al.* (1971)
14	Bis-(3-oxoundecyl)-disulfide	67–67.5	—	Roller *et al.* (1971)
15	*S*-(*trans*-3-Oxoundec-4-enyl)thioacetate	Oil	—	Roller *et al.* (1971)
16	3-Hexyl-4,5-dithiacyclo heptanone	Oil	−65	Roller *et al.* (1971)
17	(−)-Bis-3-acetoxyundec-5-enyl disulfide	—	—	Moore *et al.* (1972)
19	*S*-(−)-3-Acetoxyundec-5-enyl thioacetate	—	−25 ± 10	Moore *et al.* (1972)
20	bis(3-Oxoundecyl)tri-sulfide	60.5–61	—	Moore (1971)
21	bis(3-Oxoundecyl)tetra-sulfide	32–33	—	Moore (1971)
22	Laurencin	73–74	+70.2	Irie *et al.* (1965)
24	Laureatin	82–83	+96	Irie *et al.* (1968b)
27	Isolaureatin	83–84	+40	Irie *et al.* (1968c)
29	Ancepsenolide	91.5–92	Variable	Schmitz *et al.* (1966)
32	Hydroxyancepsenolide	122.5–123.7	Variable	Schmitz *et al.* (1969)
33	2-(13′-Carboxy-14′,15′-diacetoxyhexadecanyl)-2-penten-4-olide	81.1–82.9	−8.3	Schmitz and Lorance (1971)
34a	Prostaglandin-15-*epi*-PGA₂	—	—	Weinheimer and Spraggins (1969)
34b	Prostaglandin-15-*epi*-PGA₂ acetate methyl ester	—	—	Weinheimer and Spraggins (1969)
37	Prostaglandin-15-*trans*-PGA₂	Oil	128	Bundy *et al.* (1972a)

REFERENCES

Ali, A., Saranthakis, D., and Weinstein, B. (1971). *J. Chem. Soc. D* 940.

Boylan, D. B., and Tripp, B. W. (1971). *Nature (London)* 230, 44.

Bundy, G. L., Daniels, E. G., Lincoln, F. H., and Pike, J. E. (1972a). *J. Amer. Chem. Soc.* 94, 2124.

Bundy, G. L., Schneider, W. P., Lincoln, F. H., and Pike, J. E. (1972b). *J. Amer. Chem. Soc.* 94, 2123.

Burgstahler, A. W., and Groginsky, C. H. (1969). *Trans. Kans. Acad. Sci.* 72, 486.

Cameron, A. F., Cheung, K. K., Ferguson, G., and Robertson, J. M. (1965). *Chem. Commun.* 638.

Cameron, A. F., Cheung, K. K., Ferguson, G., and Robertson, J. M. (1969). *J. Chem. Soc. B* 559.

Ciereszko, L. S., Sifford, D. H., and Weinheimer, A. J. (1960). *Ann. N.Y. Acad. Sci.* 90, 917.

Cook, A. H., Elvidge, J. A., and Heilbron, Sir I. (1948). *Proc. Roy. Soc. B* 135, 293.

Cook, A. H., Elvidge, J. A., and Bentley, R. (1951). *Proc. Roy. Soc. B* 138, 97.

Das, K. C., and Weinstein, B. (1969). *Tetrahedron Lett.* 3459.

Fukuzawa, A., Kurosawa, E., and Irie, T. (1972). *J. Org. Chem.* 37, 680.

Harrington, G. W., and Holz, G. G. (1968). *Biochim. Biophys. Acta* 164, 137.

Heller, J. H., Heller, M. S., Springer, S. and Clark, E. (1957). *Nature (London)* 179, 919.

Hilditch, T. P., and Williams, P. N. (1964). "The Chemical Constitution of Natural Fats," 4th ed. Wiley, New York.

Hincliffe, P. R., and Riley, J. P. (1971). *J. Amer. Oil Chem. Soc.* 48, 514.

Hlubucek, J. R., Hora, J., Toube, T. P., and Weedon, B. C. L. (1970). *Tetrahedron Lett.* 5163.

Holton, R. W., Blecker, H. H., and Stevens, T. S. (1968). *Science* 160, 545.

Irie, T., Yamamoto, K., and Masamune, T. (1964). *Bull. Chem. Soc. Jap.* 37, 1053.

Irie, T., Suzuki, M., and Masamune, T. (1965). *Tetrahedron Lett.* 1091.

Irie, T., Suzuki, M., and Masamune, T. (1968a). *Tetrahedron* 24, 4193.

Irie, T., Izawa, M., and Kurosawa, E. (1968b). *Tetrahedron Lett.* 2091.

Irie, T., Izawa, M., and Kurosawa, E. (1968c). *Tetrahedron Lett.* 2735.

Irie, T., Izawa, M., and Kurosawa, E. (1970). *Tetrahedron* 26, 851.

Jaenicke, L., Akintobi, T., and Müller, D. G. (1971). *Angew. Chem. Int. Ed. Engl.* 10, 492.

Klenk, E., and Eberhagen, D. (1962). *Hoppe-Seyler's Z. Physiol. Chem.* 328, 189.

Kurosawa, E., Izawa, M., Yamamoto, K., Masamune, T., and Irie, T. (1966). *Bull. Chem. Soc. Jap.* 39, 2509.

Lambertsen, G., and Holman, R. T. (1963). *Acta Chem. Scand.* 17, 281.

Lee, R. F., and Loeblich, A. R., III (1971). *Phytochemistry* 10, 593.

Lee, R. F., Nevenzel, J. C., Paffenhöfer, G.-A., Benson, A. A., Patton, S., and Kavanagh, T. E. (1970). *Biochem. Biophys. Acta* 202, 386.

Lovern, J. A. (1964). *Annu. Rev. Oceanogr. Mar. Biol.* 2, 169.

Machlis, L., Nutting, W. H., Williams, M. W., and Rapoport, H. (1966). *Biochemistry* 5, 2147.

Machlis, L., Nutting, W. H., and Rapoport, H. (1968). *J. Amer. Chem. Soc.* 90, 1674.

Malins, D. C., and Wekell, J. C. (1970). *In* "The Chemistry of Fats and Other Lipids" (R. J. Holman, ed.), Vol. 10, pp. 339 ff. Pergamon, Oxford.

Moore, R. E. (1971). *J. Chem. Soc. D* 1168.

Moore, R. E., Pettus, J. A., Jr., and Doty, M. S. (1968). *Tetrahedron Lett.* 4787.

Moore, R. E., Mistysyn, J., and Pettus, J. A., Jr. (1972). *Chem. Commun.* 326.

Morris, L. J., Harris, R. V., Kelly, W., and James, A. T. (1968). *Biochem. J.* **109**, 673.

Müller, D. G. (1968). *Planta* **81**, 160.

Müller, D. G. (1972). *Naturwissenschaften* **59**, 166.

Müller, D. G., Jaenicke, L., Donike, M., and Akintobi, T. (1971). *Science* **171**, 815.

Ohloff, G., and Pickenhagen, W. (1969). *Helv. Chim. Acta* **52**, 880.

Parker, P. L., Van Baalen, C., and Maurer, L. (1967). *Science* **155**, 707.

Pettus, J. A., Jr., and Moore, R. E. (1970). *J. Chem. Soc. D* 1093.

Pettus, J. A., Jr., and Moore, R. E. (1971a). *Abstracts*, 161*st* National Meeting, Amer. Chem. Soc., Los Angeles, CA, March, 1971, ORGN 113.

Pettus, J. A., Jr., and Moore, R. E. (1971b). *J. Amer. Chem. Soc.* **93**, 3087.

Roller, P., Au, K., and Moore, R. E. (1971). *J. Chem. Soc. D* 503.

Schmitz, F. J., and Lorance, E. D. (1971). *J. Org. Chem.* **36**, 719.

Schmitz, F. J., Kraus, K. W., Ciereszko, L. S., Sifford, D. H., and Weinheimer, A. J. (1966). *Tetrahedron Lett.* 97.

Schmitz, F. J., Lorance, E. D., and Ciereszko, L. S. (1969). *J. Org. Chem.* **34**, 1989.

Schmitz, F. J., Lorance, E. D., and Ciereszko, L. S. (1970). *In* "Food-Drugs from the Sea" (H. W. Youngken, Jr., ed.), p. 315. Marine Technological Society, Washington, D.C.

Schneider, W. P., Hamilton, R. D., and Rhuland, L. E. (1972). *J. Amer. Chem. Soc.* **94**, 2122.

Stansky, M. E. (1967). "Fish Oils: Their Chemistry, Technology, Stability, Nutritional Properties, and Uses." Avi Publ., Westport, Connecticut.

Weinheimer, A. J., and Spraggins, R. L. (1969). *Tetrahedron Lett.* 5185.

Weinheimer, A. J., and Spraggins, R. L. (1970). *In* "Food-Drugs from the Sea" (H. W. Youngken, Jr., ed.) p. 311. Marine Technological Society, Washington, D.C.

Winters, K., Parker, P. L., and Van Baalen, C. (1969). *Science* **163**, 467.

Youngblood, W. W., Blumer, M., Guillard, R. L., and Fiore, F. (1971). *Mar. Biol.* **8**, 190.

AUTHOR INDEX

Numbers in italics refer to the pages on which the complete references are listed.

183

SUBJECT INDEX

193